The Zen in
Modern Cosmology

The Zen in Modern Cosmology

Chi-Sing Lam
McGill University
University of British Columbia

World Scientific

NEW JERSEY • LONDON • SINGAPORE • BEIJING • SHANGHAI • HONG KONG • TAIPEI • CHENNAI

Published by

World Scientific Publishing Co. Pte. Ltd.
5 Toh Tuck Link, Singapore 596224
USA office: 27 Warren Street, Suite 401-402, Hackensack, NJ 07601
UK office: 57 Shelton Street, Covent Garden, London WC2H 9HE

British Library Cataloguing-in-Publication Data
A catalogue record for this book is available from the British Library.

THE ZEN IN MODERN COSMOLOGY

ISBN-13 978-981-277-185-8
ISBN-10 981-277-185-9
ISBN-13 978-981-277-186-5 (pbk)
ISBN-10 981-277-186-7 (pbk)

Printed by FuIsland Offset Printing (S) Pte Ltd, Singapore

To Cynthia and Phoebe

Preface

Two of the most important advances in physics and astronomy in the past ten years are 1) the discovery that the expansion of our universe is speeding up, at just the right amount to render our universe spatially flat, and 2) the precise measurement of the cosmic microwave background radiation by the WMAP satellite, which supports the inflationary theory of the universe advanced by Alan Guth almost thirty years ago. No less significant is the discovery that neutrinos have a tiny mass, in the range that is completely consistent with the total amount of matter in the universe according to a theory known as leptogenesis. This also supports the inflationary theory indirectly, for without something like the leptogenesis theory to generate matter, the inflationary theory would leave behind almost no matter in the present universe.

I attempt to explain these important advances in this book. It is aimed at the layman who is curious about science, and does not assume any prior exposure to physics, mathematics, and astronomy beyond that at the high school level. More involved concepts which may require a bit more knowledge of physics and/or mathematics are relegated to an appendix of endnotes at the back of the book.

According to the theory of inflation, our universe started some 13.7 billion years ago from a tiny speck with no matter and very little energy. In other words, from a very tiny world which was almost empty. What made it grow explosively in size, in energy, and in matter content will be explained in the text, but the emptiness of the early universe also reminds me of the emptiness emphasized by Zen Buddhism. I am neither a Buddhist nor a Buddhism scholar, but I think the similarity and difference between these two notions of emptiness are amusing enough for at least a cursory comparison. This is why the first five chapters of the book contain a brief introduction to Buddhism. It is also the reason why the word Zen appears in the title of the book. The bulk of the book, however, is on the science of cosmology, and not on the philosophy of Buddhism.

The cover painting makes an attempt to depict some important features of an inflationary cosmology. The mountain is an artistic rendition of the 'mountain curve' in Fig. 42, which summarizes some important features of modern cosmology. The steep cliff on the left represents the inflationary period when an explosive growth of the universe occurs; its barrenness and the presence of the pagoda are there to remind us of the emptiness of the early universe. The gentler slope on the right of the mountain depicts the classical Big Bang periods, first dominated by radiation then by matter, where the richness of the present universe began to emerge. The altitude at any point in the mountain represents the wave number, which is inversely related to the size of a disturbance of the universe. It is this disturbance originated in the inflationary era that generates the fluctuations of the microwave radiation observed by WMAP, and acts as the seed for the galaxies and stars to be formed later. The position of the clouds across the mountain tells us roughly the observed size of these disturbances. The idea of using a mountain to describe many of the important features occurred independently to James Bjorken, who developed the connection between the mountain and many important features of cosmology much more extensively than is sketched here. The idea of the cloud is due to him. I am much indebted to him for explaining to me the other features of the connection not discussed in this book.

The beautiful painting is executed by my old friend Nora Li, to whom I owe a deep gratitude. There are many other people to thank for suggestions to improve the early drafts. Among them I should mention Bill Chan, John Crawford, Chi-Fang Chen, David Jackson, David Kiang, Jonathan and Lilian Lee, Kelvin Li, Siu-Kay Luke, Tung-Mow Yan, and my wife Cynthia. Special thanks should go to James Bjorken, Wu-ki Tung, and my daughter Phoebe, who have each read large parts of the manuscript and made detailed suggestions for changes. Without them the book would have been in much worse shape. Last but not least, I am grateful to my editor Ian Seldrup, who is skillful, patient, sympathetic, and very willing to help.

With the launch of the Planck satellite next year, and with the many other instruments both in preparation and in the design stage, much more data will be available in the coming years to vastly expand our knowledge of cosmology. I hope this book serves to prepare readers for these exciting events to come in the near future.

Chi-Sing Lam
September, 2007
Vancouver

Contents

1

Out of Emptiness

When the Fifth Patriarch of Chinese Chan (Zen) Buddhism decided to retire, he gathered his disciples and told them each to compose a stanza showing their comprehension of Buddhism. This would help him decide who should succeed him to become the next Patriarch.

The disciples all assumed the position would pass on to their leader, the head disciple Shen Xiu (神秀), so none bothered to write a stanza. Shen Xiu, realizing that the command of their teacher had to be carried out, but not wanting to appear ambitious and eager for the position, decided to write it anonymously on a blank wall. The stanza reads:

身是菩提樹，心如明鏡臺，時時勤拂拭，勿使惹塵埃

Translated, it says

Our body is a bodhi tree,
The mind a bright mirror stand.
Having them constantly wiped and cleaned,
No dust is allowed to land.

'Bodhi' is the Sanskrit word for enlightenment; for that reason the papal tree under which Buddha Sakyamuni attained his enlightenment is called a bodhi tree. In this poem, the body is likened to the sacred bodhi tree.

This anonymous stanza on the wall attracted a lot of attention. It also received high praise from the Fifth Patriarch, who told the gathering crowd to learn and practice its teaching. Privately, the Fifth Patriarch was not all that impressed by the poem, so when Shen Xiu came to him later to confess his authorship, and to ask for his comment and advice, he told Shen Xiu frankly that the stanza showed him to have reached the door of enlightenment, but not yet to have entered. Shen Xiu was then told to seek out his true mind and come back with a better stanza.

In the meantime, Shen Xiu's poem became very popular and was recited all over the temple. It came to the ears of an illiterate young worker who was not even a monk, much less a disciple of the Fifth Patriarch. Nevertheless, this young worker, whose name was Hui Neng (慧能), possessed a deep, innate understanding of Buddhism despite being illiterate. He also thought that the stanza was shallow, and decided to make one up himself to reveal what he thought Buddhism should be. Being illiterate, he had to beg somebody to inscribe it on another blank wall. It reads:

菩提本無樹，明鏡亦非臺，本來無一物，何處惹塵埃

Translated, it says

Bodhi is no tree,
Nor bright mirror a stand.
Nothing is really there,
Where can any dust land?

The Fifth Patriarch saw that and considered Hui Neng's stanza really profound, so he secretly passed on the robe and the begging bowl, which are the insignia of the patriarchship. Hui Neng thus became the Sixth and perhaps the most famous Patriarch in Chinese Zen Buddhism.

Much more of this story will be told in Chap. 4.

I first heard this tale when I was a child. It sounded fantastic and profound. Too profound! What was Hui Neng talking about? Why did he negate the existence of the bodhi tree and the mirror stand and say that "*nothing is really there*"?

When I asked my elders to explain, they merely said that the Zen philosophy was very deep, and often appeared incomprehensible to the uninitiated. For example, it would ask a question such as: "What is the sound of one hand clapping?" Well, that answer did not help me understand but I dared not ask any further. I did find out later that 'void' or 'emptiness' (sūnyatā in Sanskrit) was a very central theme in Zen Buddhism.

Decades later, I was privileged to be in the audience at the Stanford Linear Accelerator Laboratory when Alan Guth gave his first talk on his new theory of the 'Inflation of the Universe.' He called 'inflation' the 'ultimate free lunch,' because the whole complexity of our enormously large universe today all seemed to have come from a tiny one which was nearly empty, with no matter and almost no energy.

Whenever I tried to explain this to a layman, all I got was a blank stare, a shrug of the shoulders, and a universal reaction of complete disbelief. How could our big universe come from a tiny one? What made it grow? How could it be nearly empty? Where did all the stars and galaxies come from then? Everything seemed so inconceivable that it appeared to be even more mysterious than Hui Neng's stanza, and more profound than Zen.

The main purpose of this book is to answer these questions, and to explain the empirical and observational evidence in support of the assertion that the universe started out almost empty. To prepare for these explanations, we shall have to step through some relevant discoveries in physics and astronomy in the past one hundred years. Then we have to understand the inflationary mechanism which Guth calls the free lunch, a mechanism which causes the tiny universe at the beginning to grow. It grows not only in size, by a gigantic multiple, but also in its energy content, by an even more fantastic amount. The motivation and the support of the inflationary theory will be discussed in Chaps. 16 and 17.

But what led to the inflation and what happened before it? That we really do not know, though there is no shortage of theories and speculation, ranging from the superstring theory to quantum loop gravity, and many others. I shall discuss none of these in this introductory book because I have no idea which is correct.

With energy present after inflation, one might think that some of it could be converted to matter by Einstein's famous formula $E = mc^2$, and that would explain where the matter of our universe came from. Unfortunately, that does not work because energy can be used to create only an equal amount of matter and anti-matter. Yet our world is populated with lots of matter and very little or no anti-matter, so the presence of matter alone without anti-matter is a mystery that Einstein's formula cannot resolve. Unlike energy, matter has the connotation of being something permanent, something tangible, and something that you can hold in your hands. If it is 'permanent' and there is no matter in the universe to start with, how can it suddenly appear? That question will be answered in Chap. 18.

Zen and Cosmology: two profound thoughts, both advocating emptiness. Could they be related?

The Chinese phrase "本來無一物" in Hui Neng's stanza, translated above as "nothing is really there," could just as well be translated as "nothing was *originally* there". Interpreted that way, it might even seem that Hui Neng already knew about the universe being nearly empty at the beginning!

To find out whether Hui Neng really knew about the primordial universe and what is the motivation and the origin of the emptiness advocated by the Buddha and Hui Neng, a brief introduction to Buddhism is included in this book in Chaps. 2–5.

Merriam-Webster's Online Dictionary has several definitions for 'emptiness.' One is "containing nothing," and another is "lacking reality, substance, meaning." The first is more like Guth's emptiness and the second is more like Hui Neng's. These two kinds of emptiness are quite different, as one deals with the science of cosmology and the other is a philosophy of life, but there are amusing connections between the two. Hui Neng likens the emptiness of the mind to the emptiness of space in our vast universe, as both can accommodate many things. Guth goes further and tells us that the emptiness of our vast universe actually came from a tiny one with almost nothing in it. This pushes emptiness even further and in that sense modern cosmology may be considered to be even more Zen than Zen Buddhism. Yet, we might also consider both kinds of emptiness to be more of a perception than a physical reality, as we shall argue in Chaps. 16 and 20.

Let us start by trying to understand what emptiness means in Buddhism.

Figure 1: The Chinese character 'Chan' (Zen) on a starry background. Is there a similarity between the emptiness in Zen Buddhism and the emptiness of the very early universe?

2

Sakyamuni Buddha

'Zen' is the Japanese pronunciation of the Chinese character 禪, pronounced 'Chan' in Chinese. This word is taken from the Sanskrit word 'dhyāna,' meaning 'meditation.' Zen is a branch of Buddhism which emphasizes seeing deeply into the nature of things. It flourished and became a distinct school in China, later spreading to Vietnam, Korea, and Japan, and in modern times, to the rest of the world.

Given its origins, I ought to refer to it as 'Chan Buddhism,' and the title of this book should have been 'The Chan in Modern Cosmology.' Unfortunately, the word 'Chan' is much less known in the West than the word 'Zen,' so I use Zen instead of Chan everywhere.

Webster's Dictionary explains Zen as: "A Mahayana movement, introduced into China in the 6th century, A.D., and into Japan in the 12th century, the emphasis of which was upon the enlightenment for the student by the most direct possible means, accepting formal studies and observances only when they formed part of such means."

You will see in Chap. 4 that this explanation is particularly appropriate in the brand of Zen Buddhism passed down by Hui Neng.

The Concise Oxford Dictionary of World Religions explains Chan (Zen) Buddhism to be "A coalition of related ways for attaining realization, even beyond enlightenment, of the true nature underlying all appearance including one's own—and above all, that there is no duality within appearances, but only the one Buddha-nature ... Chan emerged as part of the Mahāyāna development, though naturally it traces its lineage back to the Buddha Śākyamuni..."

To understand Zen, let us now start from the beginning, find out who Buddha Sakyamuni was, what Buddhism represented, and how he invented it. In the next two chapters, the lineage and the early development of Zen Buddhism in China will be briefly discussed.

Sakyamuni was a prince in a little kingdom in present-day southern Nepal. He lived approximately from 563 BCE to 483 BCE. His name was Siddhartha Gautama, but he was commonly known as Sakyamuni, meaning the awakened one of the Sakya clan.

He led a peaceful but sheltered life in the palace, deliberately shielded from all the evils and unpleasantness of the real world by order of his father, the King. At the age of 25, he felt the call of the real world, but he was not allowed to go outside the palace. At the age of 29, the urge became irresistible, and he eventually found a way to get into the city. There he saw what is now known as the *four sights*: an old crippled man, a diseased person, a decaying corpse, and finally an ascetic.

He was shocked by this sudden revelation of human suffering from old age, disease, and death, and came to realize that these universal tragedies would befall even a prince. He was, however, impressed by the serenity and the inner calm of the ascetic, and learned from his servant that the ascetic came to be so by adhering to a strict discipline, and by renouncing all desire and hate.

The combination of shock and admiration made him decide to seek the life of an ascetic to regain his much needed inner peace. With that decision he escaped from the palace, shaved his head, changed his clothes, picked up a begging bowl, and began the life of a wandering monk searching for a teacher to guide him.

With Arada Kamala, he studied one-point concentration and deep meditation until he fully mastered the technique. Then he went on to Rudraka Ramaputra whose teaching went beyond 'nothingness' to 'neither perception nor non-perception.'

Those studies did not answer his question of how to avoid the suffering from old age, disease, and death. Without being able to find a teacher who knew the answer, he decided to seek the truth himself by going into the wilderness to ponder.

He sat under a papal tree, a kind of banyan fig tree native to India, vowing never to rise until he found the answer to his question. It is said that birds nested in his hair and spiders spun webs on his face and clothes, yet still he sat. It is also said that Mara the devil came to test him, bringing all kinds of temptation and threat, but he successfully resisted all of them.

He came to realize a truth, that the world was constantly changing, that everything was *impermanent*. He also decided that life was all suffering, for even things that appeared to provide happiness, things such as wealth, fame, power, sex, and love, were sources of sorrow when they are lost. He concluded from this observation that his most important mission was to find ways and means to alleviate the inevitable suffering in life.

One morning at the age of 35, he had figured out everything. His thought cleared up and he attained *enlightenment*. That is why the papal tree under which Buddha attained enlightenment is now known as a bodhi tree.

It became evident to him that he was not unique, that everybody and every living thing had the innate ability to attain

Figure 2: A bodhi tree.

enlightenment; it was just the society around us that confused us and prevented us from reaching that goal. In order to attain enlightenment and to relieve suffering and sorrow, we must penetrate the confusion to overcome these obstacles to enable our innate ability to see the real truth.

This innate quality is known as the *self nature*, or *Buddha nature*. It is the Buddha nature referred to in the Oxford Dictionary quotation above, a feature particularly emphasized by Hui Neng (see Chap. 5). It is also what the Fifth Patriarch thought Shen Xiu had not sufficiently unveiled in his stanza.

To summarize this line of reasoning, Sakyamuni proclaimed the *Four Noble Truths* (catvārī ārya satyāni in Sanskrit). Namely,

1. *Sorrow.* Life is suffering.
2. *Cause of Sorrow.* Suffering comes from our illusion of life, and our attachment to notions and things.

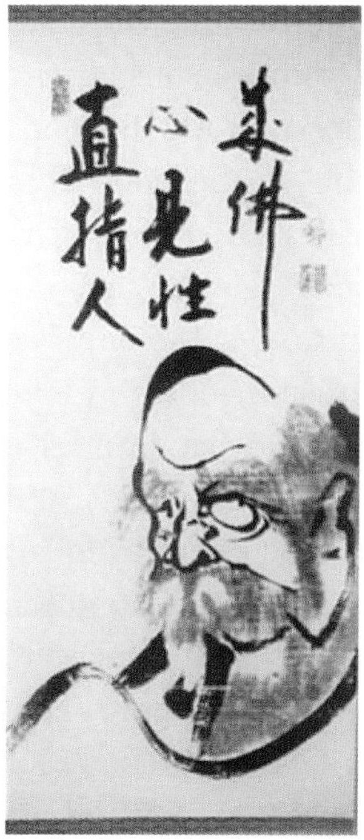

Figure 3: A painting of Buddha Nature by Hakuin Ekaku of the 18th century.

3. *Cessation of Sorrow.* This occurs when we realize that all our attachment and desire is *empty.*
4. *Path.* There are eight ways of achieving the end of suffering, known as the eightfold path (ārya astāngika mārga in Sanskrit).

I will not discuss what the eightfold path is, but let me elaborate a bit on the other points.

Two central themes of Buddhism are *impermanence*, the simple notion that things always change, and *interdependence*, the realization that things and events depend on one another and do not occur in isolation. To a physicist, these two are just the basis of any dynamics, but to a Buddhist, a deep understanding of these facts of life can lead to salvation.

Buddha asserts life to be suffering (the first *noble truth*), and sorrow to be derived from personal desires and attachments to emotions and the material world (second *noble truth*). Because of impermanence, none of the good things last forever. Their unavoidable loss ultimately generates suffering, and even attempts to keep them could cause agony. What we try to pursue is often elusive and empty of meaning. The only way to avoid sorrow is to avoid desire and over attachment (third noble truth).

Interdependence makes what belongs to me also belongs to everybody, rendering '*self*' meaningless. With this realization, one is freed from over attachment and from suffering caused by one's ego. It also makes one compassionate towards others' misfortune and suffering. This detachment makes everything around me somewhat irrelevant, as if it were *empty*. An empty mind uncluttered by mundane worries is also a great mind because it is more capable of accepting fresh ideas and seeing one's true Buddha nature.

Impermanence makes it difficult to recognize reality. Possessions and whatever appears to be real today could just be an illusion, gone tomorrow. Interdependence makes it even harder to isolate and recognize reality. Nothing is what it seems to be: nothing is real, except the innate Buddha nature within you. That is why the Buddha nature is so precious.

Various examples from daily life are often used to illustrate and strengthen these arguments, and meditation is encouraged to clarify one's thoughts. Ingenious, but sometimes complicated, arguments are often advanced to prove the correctness of these assertions.

Sakyamuni's teachings always dealt with the suffering in life and how to alleviate it. He avoided talking about God, the afterlife, and other mystical topics, because they do not solve the problem of human suffering in life by directly attacking its cause.

If Buddhism is defined by Buddha's teaching, then it is not even a religion in the sense that deities are not involved. It is just a philosophy of life. If you go to a temple or talk to a layman, you are liable to get a different impression of Buddhism, but these additional elements did not originate from Sakyamuni.

Interestingly, Confucius of China, who happened to be born about the same time as Sakyamuni and whose teaching greatly affected Chinese ethics and morality, also refused to talk about God, the devil, and the afterlife. Sometimes people in the West

Figure 4: This gigantic statue of the Buddha carved into the cliff in the Bamiyan Valley of Afghanistan was believed to be built in the fifth or the sixth century. Unfortunately, it was completely destroyed by the Taliban in 2001.

refer to Confucianism as a religion, but whatever extra trimmings that might have been added to it really had nothing to do with Confucius. In this way, the philosophies of Confucius and Sakyamuni are very similar.

Over the next two millennia, Buddhism gradually spread across much of central and east Asia. It is divided into two main schools, the Mahayana and the Theravada. The former is prominent in China, Korea, Japan, and Vietnam, and the latter is popular in Sri Lanka and Southeast Asia. Zen Buddhism is a branch of the Mahayana school.

Many impressive temples and sculptures of Buddha were created in Asia. Figures 4 and 5 show two impressive ones: a

Figure 5: The south gate of Angkor Thom, a 3 km by 3 km walled city protected by a moat in Angkor, capital of the Khmer empire, which is located at present-day Siem Reap in Cambodia.

gigantic statue in Afghanistan, built in the 5th or 6th century but destroyed by the Taliban in 2001, and one of the four Buddha faces above the south gate of the Angkor Thom in Siem Reap, Cambodia, built in the 12th and the 13th centuries.

3

A Flower and a Smile

The person who succeeded Sakyamuni was Mahakasyapa, considered the first Zen patriarch in India. This lineage of patriarchship passed down from person to person, or as Buddhism put it, from mind to mind, until the 28th patriarch Bodhidharma, who left India for China, possibly in the 5th century. He continued the lineage in China as the First Patriarch of Chinese Zen Buddhism. Hui Neng, about whom we have a lot more to say in the next chapter, is the sixth and the last Patriarch.

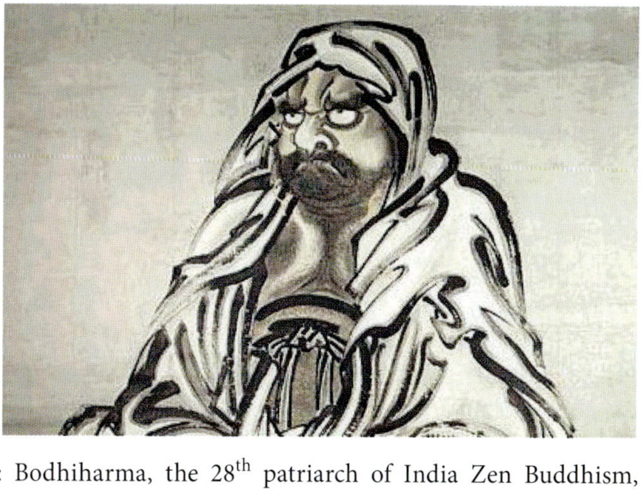

Figure 6: Bodhiharma, the 28th patriarch of India Zen Buddhism, and the First Patriarch of Chinese Zen Buddhism.

The story of transmission from Sakyamuni to Mahakasyapa is an interesting one. It illustrates an important trait of Zen Buddhism, that communication is often made directly between minds, without a word being exchanged.

Sakyamuni had always been very impressed by the insight and the practice of his disciple Mahakasyapa. One day, when Buddha was about to give a sermon on Vulture Peak, he was presented with a golden lotus flower to show a person's respect.

With the attentive audience waiting for his sermon, Buddha stood there facing them, without saying a word for a very long time. Finally, he held up the flower in front of the assembly, still not saying anything. Nobody knew what to make of it, except Mahakasyapa, who smiled.

That smile earned him the transmission.

Seeing that smile, Sakyamuni told him, "You have the treasury of the true Dharmic eye, the marvellous mind of nirvana, now I entrust it to you, Mahakasyapa." That was how the transmission to Mahakasyapa occurred.

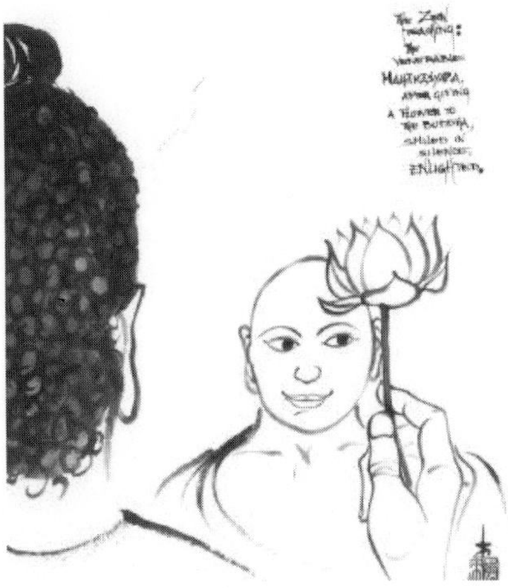

Figure 7: Mahakasyapa's smile.

Why the Buddha was silent for a long time, what the meaning of the raised flower was, and what went on in the minds of the Buddha and Mahakasyapa, presumably they knew but they did not tell anybody. The reason for the smile and what it meant is a subject of great debate for later Zen masters. Whatever the true interpretation is, this telepathic, non-verbal communication that occurred between Buddha and Mahakasyapa is a real trait of Zen that occurs over and over again. Maybe some Zen ideas can only be communicated that way!

To try to understand Zen, it is useful to bear in mind what the famous Japanese Zen scholar D. T. Suzuki said: "Zen is decidedly not a system founded upon logic and analysis… If I am asked, then, what Zen teaches, I would answer, Zen teaches nothing. Whatever teachings there are in Zen, they come out of one's own mind. We teach ourselves, Zen merely points the way."

That is the innate *self nature* at work. It certainly applies to Buddha, and to Hui Neng.

It also applies to the discovery of most original ideas on anything.

4

Hui Neng

Hui Neng was born to the poor Lu (盧) family in Guangdong Province in southern China in 638 CE, and he died in 713 CE. Hui is pronounced 'whay,' and the 'eng' in Neng is pronounced like 'unc' in the word junction. Both characters together make up his given name.

There is quite a saga associated with Hui Neng's ascension to the patriarchship, so intriguing that it is worthy of a movie.

In Chap. 1, I related Hui Neng's stanza. Being illiterate, he had to beg somebody to write it for him on a blank wall.

That poem attracted a large crowd. Some thought it insightful, but others considered it hostile to Shen Xiu, the head disciple whom they all respected. Still others ridiculed it as worthless because it was written by a lowly illiterate worker. This crowd in turn attracted the attention of the Fifth Patriarch, who came over to read the stanza. He was impressed, but upon hearing the comments all around him, he was also worried for Hui Neng's safety. To protect Hui Neng, he took off his shoe and wiped the stanza off the wall, telling the crowd: "Forget it, this stanza does not show the essence of self nature. It is better for you to practice from Shen Xiu's writings."

The next day, the Fifth Patriarch came into the room where Hui Neng was using a tilt hammer to hull the rice. The Fifth Patriarch asked Hui Neng, "Is the rice hulled?" Hui Neng replied, "Yes, it is, but not yet sifted."

That was a Zen exchange.

The Fifth Patriarch's question actually meant: "Have you recognized your innate Buddha nature?" Hui Neng's answer meant: "Yes, I have, but I still need your advice and guidance."

Without another word, the Fifth Patriarch used his stick to knock at the mortar three times and left. Nobody else in the room noticed it, but Hui Neng understood that the Fifth Patriarch wanted to see him at three *geng* (around midnight).

Once again, a non-verbal communication.

That night, the Fifth Patriarch let Hui Neng into his room, closed the door, and covered the window so nobody could look in from the outside. Without saying a word, he pulled out a copy

Figure 8: Hui Neng working in the temple.

of the Diamond Sutra (Vajracchedika-prajñāpāramitā-sūtra in Sanskrit), and started explaining its origin and its meaning, page by page. When he came to the phrase, "One should use one's mind in such a way that it will be free from any attachment," Hui Neng felt thoroughly illuminated and exclaimed, "Who would have thought that self nature is so pure, so permanent, so self-sufficient, and so embracive!"

The Fifth Patriarch was happy to hear that, knowing there and then that Hui Neng was ready, so he handed over the ceremonial robe and made Hui Neng the Sixth Patriarch.

He added, "When Bodhidharma came to China, he brought along this robe to prove to the world that what he preached came down directly from the Buddha. However, dharma must be transmitted from mind to mind, with the robe being only a symbol. It is an extraneous object that people would fight for, so from now on, it is better just to pass on Buddha's teaching, and not the robe, or else fighting will break out and you might even lose your life as a result."

That is why the formal lineage of patriarchship stopped with Hui Neng. There is no Seventh Patriarch.

He also said: "You are in great danger from the hostility of the monks here. Leave immediately, and do not preach openly for some time, or else your life might be endangered."

Hui Neng came from the Guangdong Province in the south and was not familiar with the local geography. To expedite his departure, the Fifth Patriarch personally took him across the river that night.

When they were in the boat, Hui Neng took over the oars from the Fifth Patriarch and said: "When I was ignorant, it was necessary for you the teacher to show me the way. Now that I am awakened I can guide myself across the river." That was another double-meaning utterance to tell the Fifth Patriarch that he was truly ready.

It was daylight before the Fifth Patriarch returned to the temple. He summoned Shen Xiu and told him he had given the robe to Hui Neng. He also comforted Shen Xiu by telling him that although there was only one robe, there were many ways to attain enlightenment. Hui Neng, with his innate talent, was able to see his self nature quickly, whereas Shen Xiu, with his broad knowledge actually hindering him, would take a longer time to get there, but with practice he would one day attain it as well.

Shen Xiu felt ashamed to remain to face the other monks, so he left the temple without saying goodbye. He went north, to meditate and later to preach, and was fully awakened years later. He respected the talent of Hui Neng and never held any grudge against him. He was so respected as a Zen Buddhist that at the advanced age of 96, the empress summoned him to the palace to serve. He lived to be over a hundred years old.

From then on, Chinese Zen Buddhism was split into two branches. The northern branch, headed by Shen Xiu and his disciples, advocated attaining enlightenment gradually through various means of practice. The southern branch, headed by Hui Neng and his disciples, emphasized the possibility of attaining enlightenment suddenly from one's innate Buddha nature, without having to go through the traditional practice, not even reading a single sutra. The dictionary definition of Zen Buddhism quoted in the beginning of the last chapter is that of the southern branch.

Most of the important Zen figures subsequently came from the southern branch. From the thirteenth century on, Zen passed into Japan, where it flourished. For that reason, Hui Neng is widely regarded as the actual founder of modern Zen.

Now back to the story.

Figure 9: A painting by LIANG Kai, circa early 13th century, depicting Hui Neng ripping up a sutra.

After a few days, the monks in the temple realized that the robe must have been carried away by Hui Neng. They felt that one of them, probably Shen Xiu, deserved it much more than a lowly illiterate barbarian worker from the south. In great anger, several hundred of them went off to find Hui Neng and bring back the robe.

Hui Neng continued his journey south. Two months later, when he came upon the great mountain range just north of his native province Guangdong, the monks from the temple caught up with him. He tried to run and hide, but could not shake off one particular pursuer. When that monk caught up with him and demanded the robe, Hui Neng laid it on a rock, and told the monk, "The robe is a symbol of Buddha's teaching, passed on from mind to mind. It is not something that you can obtain by force." True enough, the monk was unable to lift the robe off the rock. He became so scared and moved that he eventually helped to divert away the other monks, enabling Hui Neng to escape.

At length, Hui Neng got back to his native province Guangdong. Still sought by the pursuers, he was forced to wander and hide for fifteen years. There are many stories about those years, but I will not go into them. One day after fifteen years, he showed up at the Faxing temple in Guangzhou, the biggest city in the province that used to be known in the West as Canton. At that moment the abbot of the temple was explaining the intricacies of Buddhism to a gathered audience. He saw a flag outside fluttering in the wind, and asked his audience: "Is the wind moving, or the flag?" Various responses were given, some said it was the wind, and others said it was the flag. As they argued, a clear voice suddenly came from a corner, "Gentlemen, neither the wind nor the flag is moving. It is only your minds that are moving."

Needless to say, that was Hui Neng. Recognizing the Zen nature of the statement, the abbot started a dialog with the stranger, and eventually realized the stranger was the Sixth Patriarch, the legitimate inheritor of the robe. From there on the legitimacy of his position was publicly recognized.

It was at that late date that Hui Neng asked his hair to be shaved, finally becoming a monk at the age of thirty-nine, fifteen years after he inherited the robe and the patriarchship of Zen Buddhism.

His hair was buried under a tree. Later on, a pagoda was erected in his honor, and his hair was dug up and put into the pagoda. The Faxing temple is now called the Guangxiao temple. The pagoda can be seen in Figure 10.

Figure 10: The pagoda in Guangxiao temple where Hui Neng's hair is stored.

Figure 11: Hui Neng's mummified body in the Nanhua temple.

After a short stay in Faxing temple, he moved north to Caoxi where he died in his late seventies. His mummified body, as shown in Fig. 11, is now kept at the Nanhua temple in the city of Shaoguan in northern Guangdong province.

I will skip this and the subsequent part of the story and go directly into his teachings in the next chapter.

5

The Platform Sutra

The Platform Sutra of the Sixth Patriarch is a collection of speeches and dialogs of Hui Neng. It is the only Buddhist writing in China that is ever called a sutra, showing its importance and the respect Hui Neng commands over people's minds.

Figure 12: Two plaques in the Nanhua temple, showing side by side the cover of Hui Neng's Platform Sutra, and his portrait.

The Platform Sutra dealt with many subjects, but I shall only quote some excerpts relevant to our previous discussions.

Here are some passages regarding Buddha nature and its universality:

- Our very nature is Buddha, and apart from this nature, there is no other Buddha.
- The wisdom of enlightenment is inherent in every one of us. It is because of the delusion under which our mind works that we fail to realize it ourselves.
- You should know that as far as Buddha nature is concerned, there is no difference between an enlightened man and an ignorant one. What makes the difference is that one realizes it, while the other does not.

On the subject of emptiness, or void, he seemed to say that not only should we keep our minds detached from prejudice and worldly interference, but we should also keep them open and uncluttered to allow fruitful ideas to be taken in. Here are some quotations:

- The mind is as great and as void as space… When you hear me talk about the void, do not at once fall into the idea of vacuity… The illimitable void of the universe is capable of holding myriads of things of various shapes and forms, such as the sun, the moon, stars, mountains, rivers, worlds, springs, rivulets, bushes, woods, good men, bad men, dharma pertaining to goodness or badness… Space takes in all these, and so does the voidness of our nature.
- We say that the essence of mind is great because it embraces all things, since all things are within our nature. When we see the goodness or the badness of other people, we are not

attracted by it, nor repelled by it, nor attached to it, so that our attitude of mind is as void as space.

- The mind should be framed in such a way that it will be independent of external or internal objects, at liberty to come or to go, free from attachment and thoroughly enlightened without the least beclouding.

6

Prologue to Our Universe

In the last four chapters, we found out what *emptiness* meant in Zen Buddhism. It is an attitude and a philosophy that promise to alleviate suffering in life. The emptiness of the mind is comparable to the vastness of the universe in the Platform Sutra. Just like the empty space in our vast universe, which has no difficulty accommodating all the stars, the empty mind can easily face the happiness and sadness in life and not be perturbed by it.

Zen Buddhism says almost nothing about the physical universe, though it does emphasize its vastness, its impermanence, and the mutual dependence of events. Modern cosmology traces this vast empty space full of stars back to a time when it was very small, contained no stars, no matter, and almost no energy. In short, a primordial universe that was unimaginably more empty than the present universe which Buddhists consider to be empty. How that comes about will be the topic of discussion in the remaining chapters of this book.

It all started in 1929, when Edwin Hubble inferred our universe to have originated from a huge explosion more than ten billion years ago. Though the origin of the explosion was not understood, this 'classical big bang theory' was able to explain many cosmological phenomena. However, it also had several serious problems as pointed out by Alan Guth in 1980. He proposed a 'theory of inflation' to remedy these problems (Chap. 16), a theory which also explained the origin of the explosion. This theory launches cosmology into the modern era, and asserts that our present universe is descended from a tiny primordial one with no matter and very little energy.

But how can that be so? If the universe was tiny then what made it grow to its present gigantic size? If it began with almost nothing then where did all the matter and all the energy we see today come from?

These are non-trivial questions that will take the rest of the book to answer. What makes it grow and where the present energy comes from are discussed in Chap. 16, and where matter comes from is discussed in Chap. 18.

These questions can be meaningfully answered only when we have a clear notion of what energy is, what matter is, and in what way these two are similar and different. This fundamental knowledge of physics that takes many years to acquire is summarized in Chaps. 11–13.

Modern cosmology is based on two important discoveries of the 20th century. The first is Hubble's discovery of the expanding universe just mentioned; this topic will be discussed in Chaps. 7–9. The second is the discovery of cosmic microwave background radiation (CMB) by Penzias and Wilson, discussed in Chaps. 14 and 17.

After inflation, the history of the universe can be divided into three periods, each with a different expansion rate driven by a different dominant constituent. Constituents of the universe are discussed in Chap. 10, and the different expansion rates are discussed in Chap. 15.

There are three appendices at the end of the book. Appendix A contains the endnotes explaining advanced concepts and quantitative details of the topics discussed in the text. It may be skipped in a first reading. Appendix B collects the abbreviations and mathematical symbols used in the text for easy reference. Appendix C consists of a list of the important cosmological events after inflation and reheating.

Before embarking on our journey to explore the universe as outlined above, I will review in the rest of this chapter three fundamental facts of physics and four simple mathematical notions which will be needed in the rest of the book.

The four mathematical points are:

1. ***Power Notation.*** We deal with very large numbers in astronomy, so instead of writing out the number in full, it is customary to write the number of zeros in the exponent. Thus 1,000 is written as 10^3, two million is written as 2×10^6, 3.1 billion is written as 3.1×10^9, and four billion billion billion is written as 4×10^{27}.
2. ***Logarithm.*** Logarithm (abbreviated as log) is the inverse of power. The logarithm of the number 10^3 is 3, and the log of 10^6 is 6, etc. The number 3.1×10^9, for example, is a number between 10^9 and 10^{10}, so its logarithm is a number x between 9 and 10 so that 10^x is precisely 3.1×10^9. This number turns out to be $x = 9.491$.
3. ***Proportionality.*** I will often say that a quantity A is *proportional* to a quantity B. That simply means that the magnitude of A is directly related to the magnitude of B, or more precisely, A is equal to some constant k times B, namely, $A = kB$, where k is known as the *proportionality constant.* If I

say A is *inversely proportional* to B, then I mean A is equal to k *divided* by B, for some constant k, namely, $A = k/B$.

4. ***Vector.*** A vector is a mathematical quantity that specifies a direction and a magnitude. It can also be specified by three numbers denoting its components along three mutually orthogonal directions. For example, the velocity $\vec{v} = (v_x, v_y, v_z)$ of an airplane is a vector. It points to the direction the airplane is moving towards, and its magnitude $|\vec{v}|$ is the speed of the plane. The three components v_x, v_y, v_x could be the speed of the plane moving east, moving north, and moving up, respectively.

Now the three physics facts:

- ***Gravity.*** Everything in the universe is attracted to everything else by a *universal gravitational force* discovered by Issac Newton. The discovery is supposed to have been made by seeing an apple falling from a tree, but there is no record of this tale. In fact, the discovery was triggered by his study of planetary motions around the sun. The strength of the force is proportional to each of the two masses involved, and inversely proportional to the square of the distance separating them. It is

Figure 13: The spiral galaxy M81. Stars inside this and other galaxies are held together by their mutual gravitational forces. Courtesy of ESA/INT/DSS2.

this gravitational attraction from the earth that causes objects to fall, and water to run downhill. It is also this force that pulls the planets back towards the center as they try to get away, resulting in elliptical orbits around the sun. What holds billions and billions of stars together to form a galaxy is also this force.

There are three *other* fundamental forces in Nature: electromagnetic, nuclear, and weak forces. Between elementary particles, these other forces are much stronger than the gravitational force. Nevertheless, gravity is the only force important in the study of planetary motion and the structure of the universe, for two reasons. Firstly, astronomical objects are very massive, so although the gravitational force between two tiny elementary particles is weak, it is quite large between two very massive objects. Secondly, the electric force is repulsive between (electric) charges of the same sign and attractive between charges of opposite signs, so they tend to cancel each other out in large astronomical bodies which consist of an equal number of positive and negative charges. Neither is the magnetic force important for a similar reason, therefore the *electromagnetic force* as a whole can be neglected. In contrast, the gravitational force is always attractive. Although weak between two elementary particles, every additional particle present reinforces the force so it becomes much more important in a large body with many particles than the electromagnetic force. The two remaining forces, the *nuclear* and the *weak forces*, both operate only at very short distances, and are irrelevant at the astronomical distances that concern us in this book.

- ***Speed of light.*** The ultimate speed that any object can travel, and any information can propagate, is the speed of light, often denoted by the letter *c*. The speed is about three hundred thousand kilometers per second, which is ten million times faster than a passenger airliner. It is so fast that it is capable

Figure 14: Distant lightning.

of going around the earth seven and a half times in a second. You can get an appreciation how fast that is by looking at distant lightning: you will see the flash long before you hear the thunder.

Visible light is electromagnetic radiation with a wavelength from 4×10^{-7} to 7×10^{-7} meters. Radio, microwave, ultraviolet and infrared light, and gamma rays are examples of electromagnetic radiation of other wavelengths that are used in modern astronomical observations. They all travel with the same speed *c*.

To simplify description, I often use the word 'light' to mean electromagnetic radiation of other wavelengths as well.

In daily life, we may regard the speed of light to be infinite, but not in astronomy, where the distance is vast so that the time it takes for light to travel from one place to another is significant. For example, it takes about 500 seconds for light from the sun to reach us, whose distance is 150 million kilometers. If the sun suddenly disappeared, we would still be seeing it shining for another eight minutes.

It takes light many years to reach us from any star. If it takes exactly a year, the distance of the star in kilometers is three hundred thousand times the number of seconds in a year, which comes to almost 10^{13} kilometers. That *distance* is known as a *light year*.

The nearest star other than the sun is in the constellation Centaurus, visible in the southern hemisphere. The brightest star system in that constellation consists of the bright star Alpha Centauri, and a much fainter star Proxima Centauri. The distance of the latter, 4.26 light years, is the closest star to our solar system. The former, at 4.39 light years, is not much farther, which is the reason why it appears so bright in the sky. The second brightest star in the constellation is called Beta Centauri. A line joining Alpha and Beta Centauri points to the four kite-like stars that form the famous Southern Cross, as can be seen in Fig. 15.

A related distance unit often used in astronomy is *parsec*, or pc for short, which is 3.26 light years. A kiloparsec (kpc) and a megaparsec (Mpc) are respectively a thousand and a million parsecs.

Figure 15: The bright star on the left is Alpha Centauri; the one to its right is Beta Centauri. They point to the Southern Cross whose four stars form the apex of a diamond at the center of the photograph.

- ***Wave propagation.*** Figure 16 shows a wave emitted by the light bulb on the left, at five consecutive times indicated respectively by the colors brown, green, black, blue, and red. The height of the wave at the light bulb can be seen to go from a minimum (brown) to zero (black) to a maximum (red), then back again towards the minimum (not shown). In what follows I shall call the height the *displacement* of the wave. Zero displacement is indicated by a black horizontal line, so that the displacement of the brown curve is the negative of the displacement of the red curve. The magnitude of the maximum displacement is called the *amplitude* of the wave, so that at the light bulb position, the displacement of the red curve is the amplitude, that of the brown curve is the negative of the amplitude, and the displacement of the other curves are smaller than their amplitudes.

  The time it takes to complete a cycle like that, from brown to brown, is known as a *period* — what is explicitly shown on the diagram is half a period. The inverse of a period is known as the *frequency*; it measures number of times a given displacement appears within a given time period. The distance between two consecutive peaks at any given time is a *wavelength*; 2π divided by the wavelength is called the *wave number*.

  The wave as a whole travels to the right, as can be seen by focusing on the position of any peak at the five successive times. As shown in the diagram, each peak has moved to the right by half a wavelength after half a period. Hence,

$$\begin{aligned}(\textit{wavelength}) &= (\textit{speed of propagation}) \times (\textit{period}) \\ &= (\textit{speed of propagation})/(\textit{frequency})\end{aligned}$$

  This relation is true for any wave, light and sound, for example, though the speed of propagation may differ from one kind of wave to another. For light, the speed of propagation is of course c. For sound, we will denote its speed to be c_s.

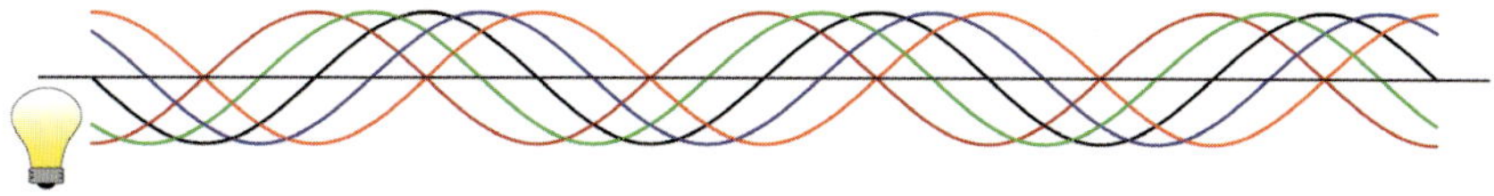

Figure 16: A wave at five different times over half a cycle. The time sequence is indicated by different colors: brown, green, and black, blue, red.

7

Does the Universe Have a Beginning?

The Bible tells us that it does: God created the universe.

Bishop James Ussher (1581–1656 CE) calculated from the Bible and other historical sources that the creation took place before nightfall preceding October 23rd, 4004 BC.

That date is a bit short, because archaeological evidence indicates that humans were present on earth long before that time, and geological evidence shows that the earth is some 4.6 billion years old.

If the universe has a beginning, a natural question to ask is: what was going on before the beginning?

When asked that question, Saint Augustine of Hippo (354–430 CE) said that God created time when He created the universe, so it is meaningless to ask what happened *before* time was created. Although he did not say so, he could have said that God also created the space to put the universe in. It is an intriguing thought that the beginning of the universe might also be the beginning of space and time.

This is a very interesting point of view. Unfortunately, I know of no theory which can implement those suggestions scientifically without invoking a deity.

What do scientists think of the origin of the universe then?

The simplest picture is to assume the universe to be eternal, because that avoids having to explain how the universe was created and what happened before the creation. Indeed, early in the 20th century, many scientists including Albert Einstein thought the universe to be eternal and static, without a beginning and an end.

That view was shattered at the end of 1920s when Edwin Hubble discovered the expansion of the universe. The universe is definitely not static then, but is it still eternal, or does it have a beginning and/or an end?

As we shall discuss, the discovery of an expanding universe *seems* to force the universe to have a beginning 13.7 billion years ago. However, nobody really knows whether this *apparent* beginning is the real beginning of the universe. We shall come back to this issue later.

Whether the universe has a beginning or not, present evidence freely extrapolated seems to indicate that there is no end. I am referring to the fact that the present universe is accelerating (a topic which will be discussed later). If it continues to do so forever, then the universe will have no end. However, we really do not know whether the acceleration will continue forever or not, so it may be a bit premature to come to such a far-reaching conclusion about its end at the present time.

Let us go back to Hubble and see how the expanding universe was discovered. It began with his discovery of external galaxies. A galaxy is a collection of many stars, bound together by the gravitational force. A galaxy often rotates about its center, sufficiently fast to flatten it into a disk-shape object like that shown in Fig. 13.

In a dark place far away from city lights, you can see a dense band of faint stars in the sky, known as the *Milky Way*. That is our own galaxy, which looks very similar to the one shown in Fig. 13. Looking out from the location of the solar system inside our galaxy, most of the stars in the disk appear to concentrate on a band, whose width is the thickness of the disk. Our galaxy is about 80,000 to 100,000 light years in diameter, about 1,000 light years in thickness, and it contains 200 to 400 billion stars.

Are there other galaxies far away from our own?

With a telescope, one can see smeared blotches of light in the sky, too diffuse to be stars. Are they external galaxies or just some interstellar clouds of dust within our own? It turns out there are both kinds, but the proof of the existence of external galaxies did not come about until the early 1920s, when Hubble found a way to measure the distance to these blotches. Some of them turned out to be too far to be inside our own galaxy. The galaxy M81 in Fig. 13, for example, is 12 million light years away, certainly way outside of our Milky Way galaxy whose radius is only a few thousand light years. As bigger and bigger telescopes are built, more and more of these galaxies show up. The following picture gives you an idea how many galaxies (most of the blotches) can be seen even in a small patch of the sky.

Hubble went on to study various properties of galaxies, including their velocities. By combining his own data and those of Vesto Slipher, another astronomer, he made an amazing discovery in 1929 that revolutionized our view of the universe.

He found that all galaxies were flying away from us, with speeds proportional to their distances, no matter in what direction a galaxy was located.

This famous discovery is known as the *Hubble Law*.

Present-day measurements put the receding velocity of galaxies to be 72 kilometers per second for a galaxy 1 megaparsec away, and twice that if it is two megaparsecs away. This number, 72 km/s/Mpc is known as the *Hubble constant*, and is usually denoted

Figure 17: There are lots of galaxies in the universe. Courtesy of NASA, ESA, S. Bechwith (STScl) and the HUDF Team.

by the symbol H_0. In mathematical language, the Hubble law can be written as $v = H_0 d$, where v and d are the velocity and the distance of the receding galaxy.

Figure 18 shows a log–log plot between the distances d measured in Mpc, and the velocity v of some nearby galaxies. Distances determined by different methods are represented by different symbols. The letter z stands for *redshift*, an important quantity which we shall discuss in Chap. 9. In the present context of small z's, it is equal to v/c, the ratio of the velocity v to the speed of light c. The solid line in the graph shows the Hubble law with a Hubble constant 72 km/s/Mpc, and the dotted lines show a 10% deviation of that on either side.

Nowadays, a similar Hubble plot is available to include galaxies much further away, out to a redshift larger than 1.7, but also with much larger error bars. In that case, the line is no longer straight, showing recent acceleration and distant deceleration of the universe that we shall return to later in this chapter.

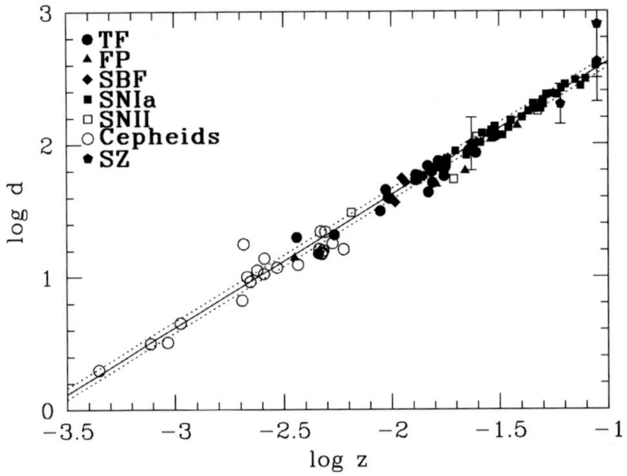

Figure 18: A Hubble plot of nearby galaxies taken from W. J. Freedman *et al.*, *Astrophysical Journal* **553** (2001) 47.

Since all the galaxies move in unison, it is more economical to regard the universe as a whole to be expanding, carrying all its galaxies along with it. If we run the time backwards using the constant Hubble velocity 72 km/s/Mpc, the universe would shrink to a point in 13.6 billion years.[1] At that time 13.6 billion years ago, the universe started from a single point (a *singularity*) and expanded isotropically in all directions, as if a big explosion had occurred to send the galaxies flying outward. This beginning of the universe is called the *Big Bang*, a name coined by the famous astronomer and science-fiction writer Fred Hoyle in the 1940s in a TV show.

Actually, the universe does not expand with a constant velocity. Both the gravitational pull from and the pressure exerted by other galaxies slow down the outward motion of any galaxy, so the expansion of the universe as a whole slows down with time. However, a phenomenon discovered in 1998 by two groups, the Supernova Cosmology Team and the High-Z Supernova Search

Team, serves to counteract this slow down. For some mysterious reason not yet completely understood, galaxies were picking up speed in the recent past rather than slowing down. The agent that causes the speedup is called *dark energy*. It is something not encountered anywhere else except perhaps at the very beginning of the universe, permitted but not required by any existing physical law, so we really have no idea what it is and where it comes from. Nevertheless, it acts to speed up the expansion of the universe. By a lucky coincidence, the total amount of slowing down due to gravity and pressure and the total amount of speeding up due to dark energy almost exactly compensate each other, resulting in an age of 13.7 billion years, very close to the 13.6 billion years obtained using a constant velocity. The actual age of the universe is therefore 13.7 billion years, not 13.6.

It is time to mention terminology I will use throughout this book. I will often discuss what happens to a galaxy in the distant past. Actually, galaxies were formed in the relatively recent history of the universe, so there were no galaxies in the distant past. Nevertheless, for simplicity of description, I will continue to use that term to refer to a small chunk of the universe at any time.

Without dark energy the expansion of the universe would slow down in time. When we run time backwards, it would shrink progressively faster so the true age of the universe becomes considerably less than 13.6 billion years. Calculation[2] shows that it takes only 2/3 of the time in that case, resulting in an age of the universe only a bit over 9 billion years. That created a crisis in astronomy a few years before the discovery of dark energy, for it was known by then that there were stars whose age were more than 11 billion years. How possibly could the universe be younger than the stars in it?

Fortunately that dilemma disappeared after dark energy was discovered, because the present age of 13.7 billion years can certainly accommodate stars that are 11 billion years old.

We move on now to a deeper question. Was the universe really created 13.7 billion years ago? If so, how was it created, and was it really born as a singularity?

13.7 billion years was the time when all galaxies piled up on one another and the universe became a single point. It therefore strongly suggests that that was the beginning of the universe, but strictly speaking it does not have to be so. It is conceivable that before that time, the (pre-)universe was present and may even have been large. For some reason at that time the whole or a part of the pre-universe came together to a point and then expanded outwards to form our present observable universe, with the rest of the pre-universe beyond our ability to see. Or else, our universe was born like the birth and growth of a rain drop in a cloud. In all these scenarios the universe as a whole was really not created 13.7 billion years ago; it is really much older than that.

In any case, our present knowledge does not allow us to extrapolate all the way back to a singularity. As the universe gets smaller, it gets hotter (Chap. 13). At some point in the distant past before we arrive at a singularity, the universe was so small and so hot that quantum effects of gravity became important. When that happens, our present ignorance does not allow us to handle the situation, so we can never extrapolate beyond that point all the way back to the singularity.

In summary, it is beyond our present ability to judge whether the Big Bang is the real beginning of the universe, and if not, what happened before it. Nevertheless, I will continue to refer to the Big Bang as the beginning of the universe, and the time since the Big Bang as the age of the universe.

8

Size and Shape of the Universe

Hubble discovered that all galaxies were flying away *from us*, isotropically in all directions. That seems to suggest that we are at the center of the universe. Are we?

A long time ago we also thought that we were at the center of the universe because the sun, the moon, and the stars all evolved around us. Galileo proclaimed otherwise, and got himself into trouble and excommunicated, but he was right. Although the moon does evolve around the earth, everything else seems to do so only because the earth rotates about its own axis.

Unless we are very egoistic, there is no reason to resurrect the notion of our central place in the universe at this late date. If not, then how can we explain the fact that all the galaxies are flying away *from us*, isotropically in all directions?

It turns out that this could be understood if we assume the universe to be homogeneous, and the galaxies to be uniformly distributed. This assumption, sometimes known as the *cosmological principle*, is consistent with our knowledge of the distribution of galaxies and the cosmic microwave background (see Chap. 14). In that case, every point in the universe is equivalent, and every one of them may be considered to be the center of the expansion.

To see how that works, imagine our universe to be a very big loaf of raisin bread, with every raisin representing a galaxy. The rise of the bread in an oven is like the expansion of the universe. No matter which galaxy (raisin) you imagine yourself to be living on, when the bread rises every other raisin is moving away from you, isotropically in all directions; so each can be thought to be the center of expansion.

Figure 19: A small loaf of raisin bread.

Now that we know the universe to be homogeneous, let us ask how large it is, whether it is flat or curved, and what possible shapes it could have.

It is hard to know how large our whole universe is because we cannot see beyond the horizon. The farthest galaxy visible to us is the one whose light emitted at the beginning of the universe 13.7 billion years ago has just reached us. If the universe were static, that galaxy is 13.7 billion light years away and that is the size of the *observable universe*, or our visible horizon. Since the universe is expanding, that galaxy would have moved away after light emission, so by now it has to be more than 13.7 billion light years from us. Calculation shows that it is some 78 billion light years away, which is then the actual size of the *observable universe* at the present. This size grows in the future and shrinks in the past. Whatever the number is, the point is that at any given time we can see only so far away, so it is hard to know how large the entire universe is.

However, with the homogeneity of the universe, there is potentially a way to find out its size provided it is finite in extent.

Before the age of airplanes and steam ships, the earth also appeared to be immeasurably vast, yet almost two thousand years ago the Greek mathematician Eratosthenes (276 to 194 BCE) found a way to determine its size by measuring the curvature of earth's surface.[1] If we can measure the curvature of the universe, then we might be able to determine its size in a similar manner. There is, however, a significant difference between the two: the surface of the earth is two-dimensional but the universe is three-dimensional. We are familiar with two-dimensional curvatures by looking down from the third dimension, but we cannot go into a fourth-dimensional space to look down at our three dimensions. Moreover, the immediate space around us is flat, so it is hard to understand what is meant by a curved three dimension, and why the universe would bother to be curved at all.

In order to build up a feeling for the real universe, let us pretend it to be two-dimensional and without a boundary. If it is flat, then it is like an infinitely large flat rubber sheet, being stretched in all directions at all times. If it has a positive curvature, then it is finite like the surface of the balloon shown in Fig. 20. The surface gets flatter as the balloon gets larger. If it has a negative curvature, then it is harder to picture, but everywhere it looks a bit like the center of a saddle, curving up one way in one direction and the opposite way in the orthogonal direction.

An animal in this universe lives on the balloon and is unaware of the presence of a third dimension. Nevertheless, assuming it to be intelligent, it can still figure out its universe to have a positive

Figure 20: A two-dimensional inflating universe.

curvature by using geometry. According to Euclidean geometry, the kind that we learn in school, the sum of the three angles of a triangle is 180 degrees. As the following example shows, this will no longer be the case for a triangle on a curved surface. If the curvature is uniformly positive, like a sphere, then the sum is more than 180 degrees. If it is uniformly negative, then the sum is less than 180 degrees.

To see why the sum of the three angles of a triangle is more than 180 degrees on a positively curved surface like a balloon, paint on it the longitude and latitude lines to make it look like the globe in Fig. 21. Consider the triangle with one apex at the north pole, and two others on the equator. The three sides of the triangle are taken to be the two longitude lines from the north pole to the two places on the equator, and a section of the equator joining the two. In this triangle, each of the two angles at the equator is 90 degrees, making their sum of the three angles 180 degrees *plus* the angle at the north pole.

Figure 21: The triangle on a globe as described in the text, with the three apexes circled in red.

Note that there are no straight lines on a balloon or any curved surface, so the best we can do for the three sides of a triangle is to take the shortest lines between the apexes. These are the proper generalization of straight lines on a curved surface because straight lines are the shortest lines between two points on a flat surface. The shortest line between two points in any dimension is called a *geodesic* line. Straight lines are geodesics on a flat surface, the longitude lines and the equator are geodesics on a globe.

Now we come to the real universe in three dimensions. The notion that such a universe may not be "flat" came from Einstein, who showed us that gravity can curl up space and space-time.[2] Curvature in a three-dimensional space can again be determined from geometry in a somewhat similar way.

Using the cosmic microwave radiation, one finds to within about 2% of possible errors that our three-dimensional universe is approximately flat. Since the allowed error is still fairly large, the universe is still allowed to have either a small positive or a small negative curvature.

If the universe is exactly flat, then it looks like an infinitely large loaf of raisin bread which continues to rise all the time. If it has a negative curvature, then it resembles a three-dimensional generalization of the middle of a saddle. It is not so easy to picture what that is so we will not attempt to do so here. If it has a positive curvature, then it looks like a three-dimensional generalization of the balloon: a 3-*sphere*.

Mathematically, the balloon is called a 2-*sphere*. Every point on its surface is equidistant from a point in the third dimension — the center of the balloon. The three-dimensional analog is a 3-*sphere*: every point on its surface is equidistant from a center point in the fourth dimension. The problem is, we do not live in four spatial dimensions, so it is hard to picture what a 3-sphere really looks like.

Other than a sphere, one might ask what other shapes the universe could look like, assuming that it has a single connected piece, has no boundary, and is finite in extent.

In the case of a two-dimensional universe, the answer is completely known. It could be a distorted sphere, obtained from a 2-sphere by pushing it in at some points and pulling it out at some other points, carefully not tearing it in the process. Mathematically, such an object is known as a *topological 2-sphere*. It could also be shaped like a donut or a distorted donut, or several of them stuck together.

The *genus* of an object is the number of independent closed curves that can be drawn on it. Two closed curves are not considered independent if one can be continuously deformed to become another. Neither is a curve considered independent if it can be continuously deformed to a point. A 2-sphere has genus 0, a donut has genus 2, and n donuts stuck together has genus $2n$.

For the real universe in three dimensions, the problem is much harder. A famous conjecture by the well-known mathematician Henri Poincaré, first put forward in 1900, subsequently modified in 1904, postulated that the only three-dimensional object with genus 0 is the *topological* 3-*sphere*. Many high-powered mathematicians have worked on this famous conjecture for the past one hundred years, but its proof did not come until very recently. For his achievement in proving the conjecture, the Russian mathematician Grigori Perelman was offered a 2006 Fields Medal. But, he declined to accept it! The Fields Medal in mathematics is like the Nobel Prize in physics, awarded to those with extraordinary achievements, but unlike the Nobel Prize, it is offered only once every four years and the recipient has to be under forty years of age. This is the first time in history that anybody has ever turned down the award.

9

Scale Factor and Redshift

Expansion of the universe puts the distance of any galaxy to us closer in the past than at the present. The ratio of these two distances is called the *scale factor*, and will be denoted by a, or $a(t)$ if we want to emphasize the time t at which it is measured. This factor does not depend on which galaxy you choose to measure[1]; it only depends on the time of that moment. By definition, it is 1 at the present time, and less than 1 in the past.

The present distance is also known as the *comoving distance.* It is the distance frozen to its current value; the actual physical distance at any time t is simply the product of $a(t)$ and the comoving distance.

The scale factor is a very important concept in cosmology. Since it changes steadily, it can be used to label time. As we shall see, it can also be used to label the temperature of the universe at a particular moment. What is more, it can be directly measured by measuring the *redshift*.

To understand redshift, let us first look at the *Doppler Effect*.

When a fast train approaches, its whistle has a higher pitch than when it is stationary, and when it moves away, its whistle has a lower pitch. This change of pitch is known as the *Doppler shift*. It happens because the sound wave is compressed as the train approaches, and stretched as it moves away. Compression results in a shorter wavelength and a higher pitch, and stretching results in a longer wavelength and a lower pitch. The faster the velocity of the train, the more the compression or expansion, and the larger the Doppler shift.

This effect is present in all waves, not just sound. For visible light, shorter wavelengths are on the blue end of the spectrum, and longer wavelengths are on the red end. The wavelength shifts towards the blue for an approaching source, hence it is known as a blue shift, and it shifts towards the red for a receding source, where it is known as a *redshift*.

The Doppler effect is used by the police to measure the speed of an automobile to catch speeding drivers. Weather bureaus equipped with Doppler radars use it to measure the velocity of a storm.

According to Hubble's observation, galaxies are flying away from us so their light experiences a redshift. The *fractional* amount of shift is a quantity also called the *redshift*, denoted almost universally by the letter z. It is defined to be the difference between the received wavelength and the original wavelength, divided by the original wavelength.

By 'original wavelength,' I mean the wavelength measured at the receding galaxy. Although we cannot go there, physics is the same there as it is on earth, so it can be measured in our laboratory from a stationary source. By 'received wavelength,' I mean the redshifted wavelength of light received on earth from the receding galaxy.

To explain how these wavelengths can be measured, let me first note that there are two types of electromagnetic radiation emitted by an atom, the *continuous spectrum*, and the *discrete spectrum*. The physical mechanism for each type will be explained in Chaps. 11 and 13.

As its name implies, the continuous spectrum consists of *all* wavelengths. It is caused by the thermal motion of atoms. This type of spectrum is useless for our purposes because there is no way to tell whether a shift has occurred in a continuous spectrum. When you see all the colors in a rainbow, there is no way you could tell whether the red color you see there is the original color or the shifted one.

The type that is useful for redshift determination is the discrete spectrum. It consists of 'spectral lines' of discrete wavelengths which characterize the atomic structure of the emitter, like fingerprints. The original discrete spectrum can be measured in the laboratory. When compared with the measured spectrum of a receding galaxy, the amount of redshift can be determined. The faster the recession, the larger the amount of shift, as is illustrated in Fig. 22.

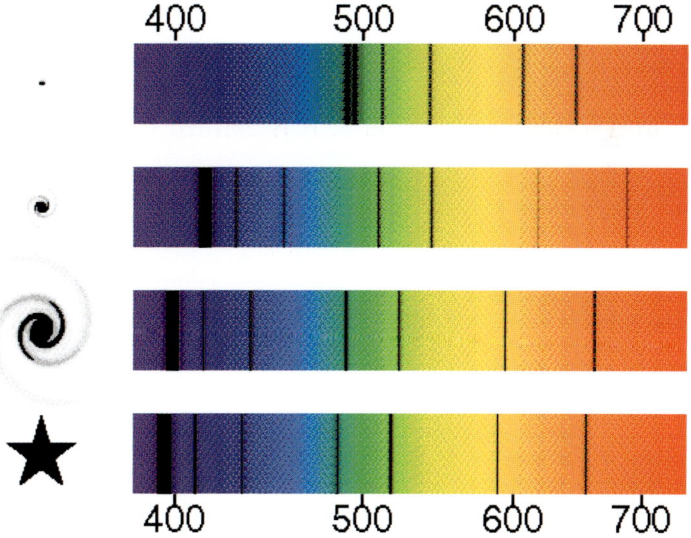

Figure 22: Discrete line spectra (black) from different galaxies, shown superimposed on the continuous spectra (colored). Their distances from us increase from bottom to top, and their red shifts also increase from bottom to top. The wavelengths on the horizontal axis are in units of 10^{-9} meters.

Now that we know how to measure it, it is time to find out what the redshift z tells us about the expanding universe.

The answer is: $z + 1$ is equal to the *inverse scale factor* at the time when light was emitted from the galaxy.[2] In other words, $z + 1 = 1/a$, where a is the scale factor.

z is small for nearby galaxies, in which case[3] it is also equal to the receding velocity v of the galaxy divided by the speed of light c, namely, $z = v/c$. This is how the velocity of recession of a galaxy is measured in the Hubble Law.

10

The Constituents in the Universe

Looking up at the sky in a place far from the city lights, especially if you look along the Milky Way or through a telescope, you will see an untold number of stars in front of your eyes. Yet there is plenty of empty space in between that you cannot quite see. For example, our sun is about 500 light seconds from us, and the nearest star outside of our solar system is about four light years away. All the space in between is mostly empty. As a result, the average density of ordinary matter that makes up a star is very low, only one atom in about 40 cubic meters.

Even so, the total amount of ordinary matter is very large because of the enormity of our universe. Knowing the size of the observable universe (about 78 billion light years), hence its volume, and the density of ordinary matter quoted above, we can multiply the last two to obtain the present amount of ordinary matter in the visible universe. It turns out about to be 7.5×10^{53} kilograms, which is equivalent to the mass of about 3.8×10^{23} suns, or 1.4×10^{80} hydrogen atoms.

The density of the universe is usually expressed as a fraction of the *critical density*. We need not know the significance of 'critical density' at the moment; it will be discussed in Chap. 15. All we have to know now is that its value is 0.97×10^{-26} kilograms per cubic meter. That corresponds to having a bit less than 6 hydrogen atoms in a cubic meter, a very small density indeed. It contains many fewer atoms per cubic meter than the best vacuum that technology on earth can provide.

The consensus now is that the average density of ordinary matter is about 4.5% of the critical density.

What may be somewhat shocking is that there is five times as much invisible matter in between the shiny stars. The invisible matter is called *dark matter.* Like ordinary matter, it is attracted to everything else by gravity, so it acts like a glue to glue stars together to form galaxies, and galaxies together to form clusters of galaxies. Dark matter has never been detected on earth, nor in cosmic rays or high energy accelerators. We recognize its existence and determine its amount through its gravitational influence. We also know that other than its gravitation, it interacts very weakly with ordinary matter and with itself, which is the reason why it is so difficult to detect.

The presence of dark matter is illustrated in the accompanying Hubble Space Telescope photograph, Fig. 23.[1] The distribution of dark matter, inferred from the way light from a distant source is bent by the gravity of the cluster, is added to the Space Telescope photograph in blue.

Figure 24[2] shows two galaxies (colored) in the cluster 1E0657-558 which collided about 100 million years ago. The dark matter content, determined in much the same way as above, is indicated by green contours. Since dark matter interacts very little, it is not retarded by the collision like ordinary matter, so dark matter shoots ahead of the ordinary matter, an effect that is clearly visible in Fig. 24.

Figure 23: A cluster of galaxies (orange and yellow), and its halo of dark matter (blue). Courtesy of R.S. Ellis (Caltech) and the Space Telescope Science Institute.

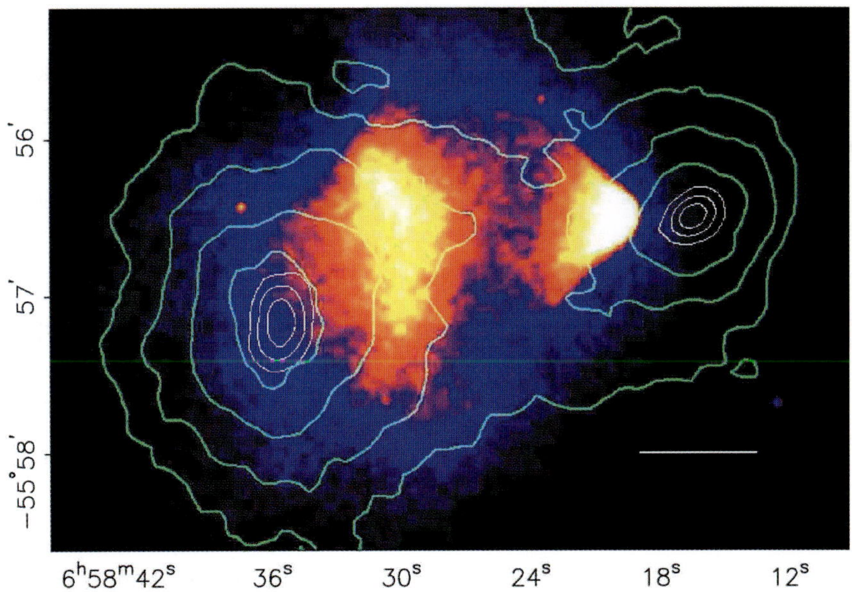

Figure 24: The dark matter (green contour) moving ahead of two colliding galaxies (red, yellow, white).

Other than ordinary matter and dark matter, there are also other kinds of objects in the present universe: photons, neutrinos, and dark energy, but we will talk about them later. In the early universe, photons and neutrinos dominated the energy of the universe, but nowadays, it is dark energy that is the most important constituent. It occupies 73% of the critical density, whereas ordinary and dark matter together constitutes only 27%.

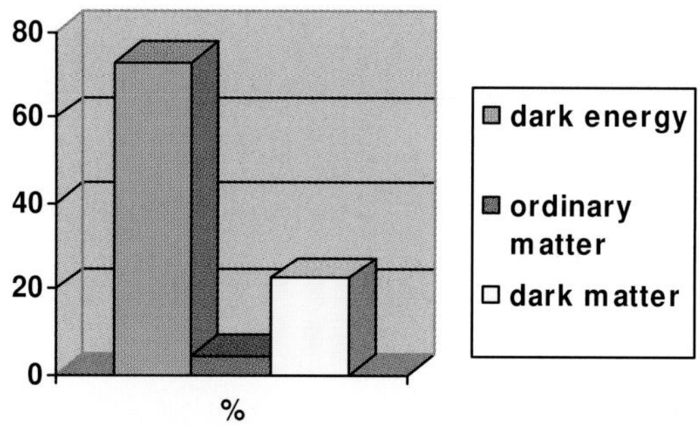

Figure 25: Constituents in our present universe. The amounts of photons and neutrinos are too small to be shown on this chart.

11

What Is Matter?

Atoms and molecules that make up planets and stars, as well as neutrinos and dark matter, are objects which I will call 'matter,' or matter particles. Photons and dark energy are examples which I will not call 'matter.' In this chapter I discuss what matter is, how many kinds of matter there are, and how its character changes with time.

In the next chapter I will discuss energy, and in what way matter and energy are equivalent. In Chap. 13, we will concentrate on a particularly important kind of energy, the thermal energy, or simply heat. We will discuss how heat affects particles and how it changes the characteristics of the universe.

The basic facts of physics learned in these three chapters will be needed for our subsequent discussions of the universe.

Now back to matter. What is it?

Matter carries the connotation of something permanent and indestructible. To a 19th century chemist, it was clear what matter was. Matter was an assembly of *molecules*, which in turn were assemblies of *atoms*. Molecules could be broken down into atoms, but atoms were indestructible and could not be broken down any further. After all, the word 'atom' came from the Greek, meaning indivisible.

There are almost a hundred different chemical elements, each characterized by its distinct atom. One might therefore say that there are almost a hundred different kinds of matter. Nineteenth century scientists thought that atoms of one kind could never be transformed into atoms of another kind, which was what alchemists tried to do but never succeeded. Atoms were thought to keep their individual identities, forever and ever.

We say a quantity is *conserved* if it does not change with time. Thus, in the 19th century, people thought that the number of atoms of each kind was conserved.

This view had to be modified in the 20th century because of two major discoveries. Early in the century, Ernest Rutherford discovered that in the process of a radioactive decay, one kind of atom could change into another kind. At first he was accused of being a modern alchemist, and he had some difficulty publishing his findings, but this momentous discovery eventually won him a Nobel Prize.

Some years after that, he also discovered that each atom possessed a very small *nucleus*. Although only about a hundred thousand times smaller than the size of an atom, it nevertheless carried almost all the atomic mass.

With these discoveries, atoms are clearly no longer indivisible, and the number of each kind of atom is no longer conserved.

By that time, *electrons* had already been discovered by Rutherford's teacher, Joseph John Thomson. Each electron carries a negative unit of electric charge, with a mass only about two thousandth the mass of a hydrogen atom.

With these ingredients, a very successful planetary model of the atom emerged, with electrons circulating a nucleus in much the same way that our planets revolve around the sun. According to this model, atoms of different chemical elements differ from one another by having different numbers of electrons revolving around different nuclei. What holds the atom together is the attractive electric force between the nucleus and the electrons. Thus, the nucleus must carry a positive electric charge, and as many units as the number of electrons, in order to render the atom electrically neutral as a whole.

Electrons in different orbits possess different amounts of energy. After the advent of quantum mechanics, it was realized that only specific orbits are allowed, and hence specific energies. These discrete energies differ from atom to atom, so their pattern can be used to identify an atom. When an electron falls from a higher orbit to a lower orbit, the released energy is carried away by the emitted light, with a wavelength related to the energy released. Thus, the discrete spectrum of wavelengths carried by the emitted light can also be used to identify an atom. This is the discrete spectrum mentioned in Chap. 9 used to measure redshifts.

The simplest atom is the hydrogen atom, with one electron going around a nucleus that carries a single unit of positive charge. The hydrogen nucleus has a name: it is known as the *proton*.

It would be simplest if a complex nucleus were just an assembly of protons, as many as the number of positive charges a nucleus is

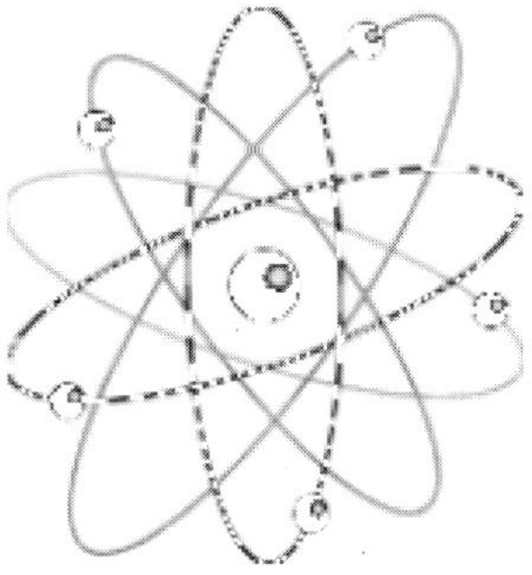

Figure 26: Planetary model of an atom, with five electrons circulating around a nucleus. The relative size of the atom and the nucleus is not correctly represented.

required to carry, and different atoms were distinguished simply by different numbers of protons and electrons.

Unfortunately, this simple picture is not correct, for a rather complicated reason.

Each proton carries a positive (electric) charge, and we know that like charges repel each other. With only electric forces between them, a bunch of protons would fly apart, rather than sticking together to form a nucleus. To be sure, there is a gravitational attraction between protons, but it is far too weak to be significant. However, there is also another attractive force operating between the protons, far stronger than the repulsive electric force, though it can be felt only at distances comparable to or smaller than the size of a typical nucleus. This force has come to be known as the *nuclear force*, or the *strong force*. With the nuclear force overpowering the repulsive electric force, protons are now expected to stick together to form a nucleus, but they still do not. Why not?

This is where *quantum mechanical effects* come in. For microscopic systems like molecules, atoms, and nuclei, we must take quantum mechanics into account. In the present context, there are two important quantum mechanical effects; both tend to weaken the effective attraction between the protons, so much so that they can no longer stick together to form a nucleus.

The first is the *uncertainty principle*, which always weakens the effective attraction at short distances (see the topic *Quantum Fluctuation* in Chap. 17 for further details). Since the nuclear force is short-ranged, this effect is quite important. The second is the *Pauli exclusion principle*, which tells us that if we put several protons into the nucleus, one after another, then those that come in later will receive progressively smaller and smaller effective attractions. It is a combination of these two effects that weakens the strong attraction between protons so much that, by themselves, it becomes impossible for them to bind together to form a nucleus.

Both of these effects apply not only to protons, but to other matter particles as well.

Even if the protons could stick together, it would still not be a good model for the atomic nucleus because it could never explain the outcome of radioactive *beta decay.* In radioactive beta decay, a chemical element of one kind is changed into a chemical element of another kind, but the masses of the corresponding atoms are very similar, differing from each other by much less than the mass of a proton. In the pure-proton model of the atomic nucleus, the only way a chemical element can change into another is for the nucleus to eject, or to absorb, one or several protons. At the same time, the atom also has to discard, or take up, the same number of electrons to remain neutral. When that happens, the mass of the atoms before and after would differ by at least one unit of proton mass (remember the electron mass is negligible compared to the proton mass), contrary to observation.

The correct view of the atomic nucleus emerged in the early 1930s, after the discovery of *neutrons* by James Chadwick. A neutron is a particle carrying no electric charge, hence the name, but otherwise very similar to the proton. It has a mass slightly larger than the proton, and like protons, there is a short-ranged nuclear force operating between neutrons, and also a similar nuclear force between protons and neutrons.

When left alone, a neutron cannot live very long. In 15 minutes or so, it will decay into a proton, an electron, and an *anti-neutrino.* This process is also known as beta *decay* because, as we shall see, it is this same process that causes the nuclei to undergo their beta decays.

I mentioned above the 'anti-neutrino.' This is a particle that I will come back to later.

With the participation of neutrons, complicated nuclei can now be formed from a mixture of protons and neutrons. The presence of neutrons dilutes the electric repulsion between protons, and at the same time it increases the strong attractions inside the nucleus. As mentioned before, the Pauli exclusion principle operates between neutrons just like it does between protons, but it does not operate between protons and neutrons because the Pauli exclusion principle only operates between identical particles. As a result, if we ignore the weaker electric repulsion between protons for the moment, the effective attraction among six protons and six neutrons is much stronger than that between 12 protons or 12 neutrons. Protons or neutrons alone are unbound, but the six protons and six neutrons stick together to form a carbon nucleus shown in Fig. 27. Similarly, two protons and two neutrons form a helium nucleus shown in Fig. 28, whereas four protons or four neutrons are unbound.

As mentioned before, the number of protons in a nucleus is equal to the number of electrons in the atom, rendering the atom

Figure 27: A carbon nucleus C^{12} with six protons and six neutrons.

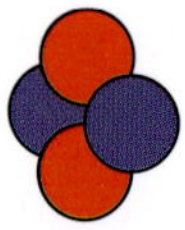

Figure 28: A helium nucleus He^4 with two protons and two neutrons. This nucleus is very tightly bound, sometimes emitted as a whole in radioactive decays. In that context, it is known as an alpha particle.

electrically neutral. In other words, the number differs from chemical element to chemical element. For a given element, the number of neutrons is however not fixed, and can often take on a range of values. Atoms with the same number of protons but different numbers of neutrons are called *isotopes* of the same element. Three isotopes of hydrogen are shown in Fig. 29. The one in the middle, with one neutron, is known as a *deuteron*, and the one on the right, with two neutrons, is known as a *triton*. The corresponding atoms are known as *deuterium* and *tritium*, respectively.

Figure 29: Three isotope nuclei of hydrogen. Proton is in red and neutron in blue.

Although the number of neutrons in a nucleus is not fixed, there cannot be too few of them or else the nucleus will not stick together. There cannot be too many of them or else the nucleus becomes unstable against beta decay.

I said previously that a neutron spontaneous decays into a proton plus something else when left alone. Why don't the neutrons inside the nucleus decay then?

Actually, sometimes they do, sometimes they do no: it's all a matter of energy. If you replace a neutron inside a nucleus by a proton, and the resulting nucleus becomes more tightly bound,[1] and sticks better together, then a neutron in the original nucleus takes advantage of this added stability and turns itself into a proton, thus giving rise to a beta decay. If the new nucleus is less tightly bound, then the neutron prefers not to do so, and the original nucleus is stable.

If there are very few neutrons in the nucleus, the Pauli exclusion principle does not favor turning a neutron into a proton, because there are already so many protons around. So, the newly turned proton can receive only a weaker attraction, which does not pay especially because there is an added repulsion. In that case, the nucleus is stable. However, if there are too many neutrons in the nucleus, then the same Pauli principle might favor turning a neutron into a proton, in spite of the added repulsion, and such a nucleus is unstable and undergoes a beta decay.

For a nucleus with too many *protons*, there is no way you can add enough neutrons to overcome the electric repulsion between protons to make everything stick together to form a nucleus. This is because after a while, those added neutrons receive so little attraction that they essentially do not count. That is why elements with very high atomic numbers (i.e. a large number of protons or electrons) do not exist, and there are only about a hundred different kinds of elements in the world.

This settles the structure of the nucleus. It is now time to explain what an anti-neutrino is.

For the first 30 years or so in the 20th century, an energy crisis persisted in atomic physics. I am not referring to a shortage of gasoline, but to the energy that was seemingly lost in beta decays. What one could detect coming *out* of a beta radioactive nucleus was a single electron. For a decay of this kind, energy conservation and Einstein's famous relation $E = mc^2$ demand that the emitted electron carries a definite amount of energy determined by the masses of the mother and daughter nuclei. Instead, the observed energy of the electron varied from event to event, and was always smaller than the predicted value. Is energy not conserved in beta decays then? Somebody as famous as Niels Bohr, the inventor of the planetary model of the atom, thought that might be the case.

In a bold move, Wolfgang Pauli, the same Pauli of the exclusion principle, resurrected the profound principle of energy conservation by postulating the existence of an unseen particle coming out of the decay, carrying away the missing amount of energy. Depending on how much energy this unseen particle carries, which varies from event to event, the electron will show up with a different amount of energy, as observed. That particle is the anti-neutrino.

To be 'unseen' the anti-neutrino has to be electrically neutral, and must not participate in nuclear interactions, so that it practically does not interact with the surrounding matter for a detector to render it observable.

After postulating this particle, Pauli exclaimed, "I have done something terrible. I have invented a particle that cannot be seen." With modern technology and large detectors, this particle has been detected since 1953. We will have more to say about it in Chap. 18.

'Neutrino' in Italian means the tiny neutral particle. It is tiny not only because it was not seen, but until quite recently it was also thought to be strictly massless. That explains why it is called neutrino, but why the prefix 'anti-'? Because for reasons to be discussed later, it is better to think of it as anti-matter, not matter.

That brings us to the discussion of *anti-matter*. Its existence was first predicted theoretically by Paul Dirac.

There are three fundamental theories of physics discovered in the first quarter of the 20th century. They are: quantum mechanics, Einstein's special relativity, and his general relativity. Quantum mechanics is needed to deal with microscopic systems like molecules, atoms, and nuclei. Special relativity is required to study objects traveling at a speed close to the speed of light. General relativity is a theory of gravity, telling us how the Newtonian gravitational law should be modified in the presence of a strong gravity.

Around 1930, Paul Dirac tried to construct a quantum theory of free electrons by taking special relativity into account. He discovered that the existence of another particle was required before his theory made sense. That new particle has to have the same mass as the electron, but carries the opposite electric charge. Moreover, when it encounters an electron, they may annihilate each other leaving only the equivalent amount of energy behind.

Since it carries a positive electric charge, this new particle is known as a *positron*.

Since it can annihilate an electron, it is also known as an *anti-electron*.

We may also call an electron an 'anti-positron.' All that Dirac's theory predicts is that there must be a pair of particles, with the same mass and opposite electric charges. It is a matter of convention to choose which of the pair to be called a particle, which the anti-particle. By convention, the stable particles that are usually found on earth, in this case the electron, is called the particle, and the positron is then called the anti-particle. An anti–anti-particle is just the particle itself.

Positrons had not been seen at the time of Dirac's theory. However, shortly afterwards, it was discovered in a cosmic ray experiment. This was a great triumph for Dirac, and a great success story for theoretical physics.

The theory of Dirac's is equally applicable to protons, neutrons, anti-neutrinos, and other elementary particles that constitute matter. It predicts the existence of anti-protons, anti-neutrons, and neutrinos. These particles have now all been found.

With these discoveries, matter consisting of protons, neutrons, electrons, and anti-neutrinos is no longer permanent, because it can be annihilated by anti-matter, and because a neutron can change into a proton in beta decay. However, as I shall explain below, the amount of matter minus the amount of anti-matter is still conserved.

Protons and neutrons are both constituents of nuclei, and they are collectively known as *nucleons*. Nucleons are the lightest of a class of heavy elementary particles called *baryons*. For most of this book, except Chap. 18, baryons and nucleons are treated synonymously because the other baryons never appear.

A neutron cannot disappear unless it beta-decays into a proton, or is annihilated by an anti-neutron. In either case, the total number of nucleons minus the total number of anti-nucleons does not change. This number is known as the *nucleonic number*, or the *baryonic number*. It is conserved.

Similarly, electrons and neutrinos are both referred to as *leptons*, and the number of leptons minus the number of anti-leptons is known as the *leptonic number*. The *lepton number is also conserved*. For example, in the beta decay of a neutron, it starts out with no leptons, and hence a zero leptonic number, and it ends up with an electron and an anti-neutrino. So, the final lepton number is still zero.

It is for the sake of keeping lepton number a conserved quantity that we call the beta-decay product of the neutron an anti-neutrino, rather than a neutrino.

There are many other matter particles, with permanence in a similar sense as explained above, but most of them are unstable and do not occur naturally on earth. Those baryons that are not nucleons are such examples. They are not as important for our purposes as the particles we have mentioned, so I will not bother you with them now.

What I call matter particles here are officially known as *fermions*, but in keeping with the historical concept of matter, it is perhaps more intuitive to refer to them as matter particles.

What about dark matter particles? They have not been *directly* detected on earth so we do not know what kind of fermions they are. What we do know from astronomical evidence is that they must be electrically neutral, stable, and interact only weakly with nucleons and electrons, if at all, for otherwise they would have been seen. We also know that they must be fairly heavy, but beyond that we know very little.

There is some circumstantial evidence suggesting that a new zoo of particles called *supersymmetric particles* may exist in nature. If such particles do exist, then many people think that the dark matter particles may be one of these supersymmetric particles. Since the evidence is no more than circumstantial, we will have to wait for more experimental studies before we can make any definitive statement on that.

There are also non-matter particles, officially known as *bosons*, whose numbers are not subject to any separate conservation law *per se*. The best known example of a boson is the *photon*, the carrier of electromagnetic waves such as light, radiowaves, microwaves, and gamma rays. If you turn on the light, photons are produced. If they strike a black wall, they are absorbed, so their number is obviously not conserved.

A photon has no mass and no charge. It always moves with the speed of light, at about three hundred thousand kilometers a second.

Matter and non-matter particles can also be distinguished by the *spins* they carry.

In addition to having a definite charge and a definite mass, a particle also carries another attribute called *spin*. Just like the earth rotates about its own axis, making a complete revolution every 24 hours, particles also rotate about their own axes. The direction of the axis and the amount of rotation defines a vector quantity (Chap. 6) called the *spin* of the particle. Unlike the earth whose axis points towards the North Star, the axes of rotation of particles can point to any direction in space. Quantum mechanics dictates that in an appropriate unit, the spin of a particle must have a magnitude specified as either a positive integer like 0, 1, 2, or a positive odd integer divided by two, like 1/2, 3/2, 5/2. These integers or half (odd) integers are also referred to as the *spin* of the particle. The matter particles we have encountered so far, proton, neutron, electron, neutrino, and their anti-particles, all have spin 1/2. Photons have spin 1. There is, in fact, a general *spin-statistics theorem* which assures us that integer-spin particles are bosons and half-(odd)-integer-spin particles are fermions.

Spin is a vector, and hence it has three components (Chap. 6), specifying how much the particle is spinning around three mutually orthogonal directions. Quantum mechanics tells us that you can only specify one of these components, not all three simultaneously. We usually choose this to be the component along its direction of motion for the reason I will discuss later. This component is called the *helicity*. Depending on the angle the spin axis makes with the direction of motion, the helicity can take on different values, positive and negative. Again quantum mechanics tells us that the angle is not arbitrary: for a spin s particle, its helicity is allowed to have only $2s + 1$ discrete values, ranging from s to $s - 1$ to $s - 2$, etc., finally down to $-s$. For particles of spin $s = 1/2$, the allowed helicities of +1/2 and −1/2 are respectively called the *right-handed* and *left-handed* helicities (see Fig. 30).

The red arrow in Fig. 30 indicates the direction of motion of a particle, and the screw tells us its spin rotation about that direction. If the screw is turned to tighten, then it moves down as indicated by the double blue arrow with the letter R, and the

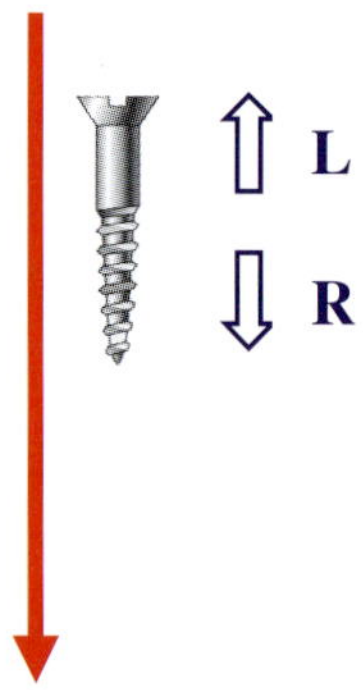

Figure 30: A diagram showing right-handed (R) and left handed (L) helicities. See the text for an explanation.

helicity is right-handed. If it is turned to loosen, then it moves up as indicated by the double blue arrow with the letter L, and the helicity is left-handed.

If the particle of spin s has a mass, it must possess all $2s + 1$ helicities. This is because special relativity tells us that physics must be the same whether we observe it at rest, or moving with a constant velocity. For example, suppose a spin-1/2 particle is right-handed to a stationary observer, namely, the screw in Fig. 30 is turned to tighten. Consider another observer who is moving in the same direction but faster than the particle, then to this observer the particle would appear to be moving backwards (or upwards in the case of Fig. 30), but the screw still turns in the same direction; hence, to this observer the particle has a left-handed helicity. In order for both observers to be describing the same physics, the particle must have a left-handed as well as a right-handed helicity. Similar arguments applied to particles with any spin s show that it must have all $2s + 1$ helicities.

This argument is no longer valid if the particle is massless, because then it moves at the speed of light so no observer can move faster to get past with it. In that case the particle may have as little as just one helicity. Photons are massless, spin $s = 1$, and hence they may not carry all the three helicities. Actually, they have only two helicities, ±1, known respectively as the right-handed and left-handed *polarizations.* The existence of both helicities for photons (rather than just a single one) is related to the conservation of parity in electromagnetic interactions (Chap. 18).

For a long time neutrinos were thought to be massless, for two reasons. The first is that no *direct* measurement has ever been able to detect a mass, even today. The second is that when it appears in beta decay, only the left-handed neutrinos and right-handed anti-neutrinos appear. With the absence of the right-handed neutrino and the left-handed anti-neutrino, it is natural to assume the neutrinos to be massless for the reason given in the paragraph before last. However, recent experiments have convinced us that at least some neutrinos do have a small mass, in which case the right-handed neutrino and the left-handed anti-neutrino must exist. We will return to that point in Chap. 18 for a detailed discussion.

Let me now discuss why we chose the direction of motion but not some other direction to label the spin component. This has to do with the conservation of *total angular momentum.* The spin of a particle is a form of angular momentum — its value while at rest.

Consider an isolated system of particles. These particles may interact with one another, but not with anything else — that is what I mean by 'isolated.' The motion of each particle at a given time is labeled by its velocity vector, but a better way to specify its motion is by its *momentum vector,* defined to be the mass of the particle times its velocity vector provided the particle is moving much slower than light speed. The reason why momentum is superior to velocity is because the total momentum of the system, obtained by adding the momenta of every particle in the assembly, is conserved.

Just as momentum is a vector concocted for linear motion so that the total momentum is conserved, angular momentum is a vector similarly defined for 'rotational' motion so that the total angular momentum[2] of an isolated system is conserved. The angular momentum of a particle consists of two parts: the orbital angular momentum and its spin. It turns out that orbital angular momentum has no component along its direction of motion,[2] and hence in the case when all the particles in the system move in one direction, the sum of their helicities are conserved. We will have an occasion to use this property in Fig. 48 of Chap. 18.

12

Different Kinds of Energy

Energy comes in many forms. Under suitable circumstances, energy of one form can change into energy of another form, but the total amount of energy remains the same. In other words, *total energy is conserved.*

That is a sacred principle in physics. As related in the last chapter, at one point it was thought by some not to be applicable in beta decay, but the invention of the anti-neutrino by Pauli restored its validity even for that process.

In this chapter we shall examine several different forms of energy: kinetic energy, potential energy, rest energy, nuclear energy, dark energy, and vacuum energy. Thermal energy will be discussed in the next chapter.

Kinetic Energy

This is the energy carried by an object in motion. The faster the object moves, the more kinetic energy it carries. The larger the mass is, the more kinetic energy it carries too. In a head-on car collision, it is this energy that does all the damage.

If m is the mass of an object moving with a speed v, then its kinetic energy is $KE = mv^2/2$, *provided* v is much smaller than the speed of light c.

This rule does not apply to a photon, because it has no mass and it always moves with the speed of light. In that case, the energy it carries is inversely proportional to the wavelength of the electromagnetic wave it resides in. Ultraviolet light has a shorter wavelength than visible light, X-rays a shorter wavelength still, and gamma-rays even shorter. Thus, a gamma-ray photon carries more energy than an X-ray photon, which in turn carries more energy than an ultraviolet photon, and a visible photon. This is why generally these shorter wavelength photons cause more damage to our body than longer wavelength ones. This is also why you merely put on some sun-tan lotion to block ultraviolet light, but you need metal or concrete protection against X-rays and gamma-rays.

Potential Energy

This is the energy stored between two (or more) particles, made possible by their mutual interactions. Its magnitude depends on their separation, as well as their individual attributes like mass and electric charge. Unlike the kinetic energy, which is always positive, potential energies can be positive or negative.

A potential acts like a bank account of energy. You can put in energy to raise the potential, and you can withdraw energy to lower the potential, all by appropriately adjusting the separation of the two particles.

The potential that one particle experiences in the presence of many other particles is the sum of the potential energies between this particle and all the other particles.

If two particles are bound together into a stable system, the energy needed to tear them apart is called their *binding energy*. We say that a system is tightly bound if its binding energy is large.

Binding energy is the amount of energy released when the two particles snap together to form a stable system.

Potential energy can be turned into kinetic energy, and *vice versa*. Since total energy is conserved, a gain in kinetic energy must be associated with a corresponding loss of potential energy. In the bank account analogy, kinetic energy is like money in hand, and potential energy is like money in the account. A negative potential energy corresponds to a negative balance in the account, at which point you are really borrowing money from the bank. Fortunately, this particular bank never charges a fee nor interest for your overdraft.

Potential energies are associated with *forces*. This comes about because there is a tendency for a potential to seek a lower value by adjusting the separation, much like water flows towards lower ground. The tendency that causes particles to change their separation is interpreted as a force. If the potential is lowered by increasing the separation, then particles appear to be pushed apart, so the force is repulsive; if the potential is lowered by reducing the separation, the force is attractive.

The potential between two particles is usually set to be zero when they are very far apart, and hence non-interacting. In that case, an attractive force corresponds to a negative potential at finite separation, and a repulsive force corresponds to a positive potential at finite separation.

The gravitational force was the first fundamental force to be discovered, by Issac Newton. As related in Chap. 6, the gravitational force between two objects of masses M and m, separated by a distance r, is GMm/r^2 (Chap. 6), where G is the Newtonian gravitational constant. Given that, it can be shown that the corresponding gravitational potential is $-GMm/r$. The minus sign comes about because gravitational force is always attractive.

Figure 31: A cliff diver.

G is a very small number, so the potential energy between two persons or two ordinary objects is negligible. However, if one of the two masses is very large, for example, that of the earth, then the potential energy can be sizable. This is illustrated in Fig. 31, where the gravitational potential energy is being converted into kinetic energy.

Rest Energy

Einstein's famous formula

$$E = mc^2$$

tells us that there is an amount of energy E equal to the product of the mass m and the square of the speed of light c stored in every particle at rest.

The speed of light c is a very large number, equal to 300,000 kilometers per *second*. Because it is so large, the amount of energy contained in any object we are familiar with is simply horrendously large. Take, for example, a kilogram of garbage. The amount of energy computed from this formula is about 30 billion kilowatt-hours, or the total amount of power produced by the Hoover Dam in two years.

Clearly, that is a lot of energy! How nice it would be if we could convert all the garbage in the world into clean energy. It would help the environment, and we would never have to fight for energy again. Many wars might thereby be avoided.

Unfortunately, to extract that energy you have to — literally — get rid of the garbage from this universe. Hiding it in a landfill or burying it at sea does not count. The permanence of matter, or in the language we learned in the last chapter, the conservation of nucleonic and leptonic numbers, tells us that the garbage cannot just vanish — unless you can find a kilogram of anti-garbage to annihilate it.

In other words, although the rest energy is there, there is no *easy* way to extract it because of the permanence of matter, or because of the lack of naturally occurring anti-matter in the world.

Maybe it is just as well that the rest energy cannot be extracted easily. Otherwise, in the wrong hands, cheap energy could be used to make bigger bombs to annihilate mankind.

The natural lack of anti-matter raises a very interesting question. Dirac's theory is symmetrical between particles and anti-particles, so how come this universe contains so few anti-particles? This is actually a deep question, which will be discussed in Chap. 18.

The problem becomes even more acute when we realize that the universe started with no matter, namely, zero baryonic and leptonic numbers. The symmetry between particles and anti-particles inherent in Dirac's theory was then obeyed. How come matter is present without an equal amount of anti-matter at the present?

Although anti-particles do not occur naturally on earth, they can be produced in high energy accelerators. To do so we need to supply at least twice the electron rest energy mc^2 to produce an electron and positron pair. Pair creations of this kind are done all the time in present day high energy accelerators. Note that it is impossible to produce an electron without a positron, because of the conservations of electric charge and leptonic numbers.

Any number of particles can be produced in these accelerators as long as energy is available and conservation laws are not violated. To be able to provide a large amount of energy, these accelerators must be very large. Figure 32 shows the size of the Large Hadron Collider (LHC), which can produce two colliding proton beams of about 7 TeV each. It is the largest machine currently on earth, scheduled to commence operation in 2007 or 2008.

Figure 32: The large circle shows the location of the underground Large Hadron Collider, straddling Switzerland (bottom) and France (top). The dotted line is the national boundary, and the long white object in the foreground is Geneva airport.

Speaking of energy, there are many units that can be used to quantify it. When atoms and nuclei are involved, it is common to use the unit of *electronvolt* (eV), which is the energy gained when a particle with one unit of electron charge slides down a one volt electrical potential. A thousand eV is known as a keV, a million an MeV, a billion a GeV, and a trillion a TeV. At the other end of the scale, a thousandth of an eV is known as an meV.

One often uses this unit to measure mass as well. By saying a mass is so many electronvolts, what one really means is that the rest energy of that mass is so many electronvolts. In this unit, the mass of an electron is 0.51 MeV, the mass of a proton is 938.3 MeV, and the mass of a neutron is 939.6 MeV.

In kilograms, the mass of a proton is 1.67×10^{-27} kilograms. Thus, in this language, one kilogram is equivalent to 938.3 divided by 1.67×10^{-27}, namely, 5.62×10^{29} MeV.

We may sometimes express energy in kilograms as well, according to this conversion relation.

So far we have talked about the rest energy of a massive particle at rest. What if it is moving? Then its energy consists of the sum of rest energy and kinetic energy. Its kinetic energy increases with the velocity of the particle, but it is no longer given by the formula $KE = mv^2/2$ when the velocity v approaches the speed of light. Instead, its kinetic energy approaches infinity in that limit.[1] For that reason it requires an infinite amount of energy to push a massive particle to the speed of light, which is an impossible task. This is why no massive particle can travel at or beyond the speed of light.

When the velocity is sufficiently large, the kinetic energy becomes much larger than the rest energy, so in comparison the latter can be neglected. A particle with that kind of velocity is known as a *relativistic* particle. Otherwise, the particle is *non-relativistic.*

Another form of energy very important both in daily life and in cosmology, associated with a large number of particles, is the *heat energy*, or *thermal energy*. We shall take that up in the next chapter.

Nuclear Energy

I was careful to say earlier that there was no *easy* way to extract the rest energy, but I did not say it was impossible. Nature has found a way to do so, and in fact, our lives depend on it. Most of the energy on earth is derived from the sun, and the energy of the sun is extracted from the rest energy difference between a helium nucleus and four protons.

When four protons turn into a helium nucleus, which has two protons and two neutrons tightly bound together (see Fig. 28), in order to conserve electric charge and leptonic number, two positrons, two neutrinos, and possibly some photons must also be emitted. The energy release from the total rest energy difference between the final product and the initial four protons is 26.7 MeV, which is the energy that can be potentially gained from each of these *fusion reactions*. Since many such reactions are going on at the center of the sun, a huge amount of energy is released to enable the sun to shine so brightly.

In case you had not realized it, that is a huge amount of energy per reaction. All other types of fuel derive their energy from the chemical reactions of atoms and molecules. Since nuclear binding energy is of the order of MeV and atomic binding energy is of the order of eV, per reaction a nuclear process is roughly a million times more energetic than a chemical process. This is why *nuclear energy* is so powerful.

If you have a cylinder of hydrogen gas, or a can of protons, there is no danger fusion would occur to cause an explosion. The electric repulsion between protons makes it very difficult to squeeze them together into a helium nucleus. Nature is able to do so in stars because their centers are very hot, thus providing a large amount of thermal energy to bang the protons into one another at high speed (see the next chapter about temperature and thermal energy). It is this high speed collision that overcomes the electric repulsion of protons to make them fuse.

The thermal energy at the center of the sun or stars is in turn derived from the large amount of gravitational potential energy released when the material in a star settles to its center. If the star were only the size of the planet Jupiter, there would not be enough gravitational energy released to allow fusion to occur. This is why the sun shines and Jupiter doesn't.

Fusion is a good thing: it provides life, as well as the beauties of sunset and starry nights. It is also a bad thing, because we have learned to use it to make hydrogen bombs. As has been said so many times, science is neither good nor bad; it all depends on whether the right persons are using it for the right purposes.

I would prefer to remember fusion by a beautiful sunset, rather than a huge mushroom cloud (Fig. 33).

Figure 33: Sunset in Vancouver.

Dark Energy and Vacuum Energy

The universe also contains lots of *dark energy* (Chap. 10), as much as 73% of its total energy. It is the dark energy that causes the present universe to accelerate (Chap. 7), and it is also this amount of dark energy that renders the universe nearly flat (Chaps. 8 and 15). It is clearly important, but what is it, and why is it there?

Unfortunately, nobody knows for sure. One possibility is that it is a *vacuum energy*. I shall presently explain what this is.

A *vacuum* is defined to be the state of lowest energy. Since particles always carry some energy, even those at rest carry a rest energy, it might seem that the lowest-energy state is just the state with nothing in it, hence the name *vacuum*. However, because of the quantum mechanical *uncertainty principle*, there can never be a steady state with no particles in it unless the particles do not interact, so strictly speaking that is not what a vacuum is.

We encountered the uncertainty principle in Chap. 11 in another guise, where it was invoked to explain the reduction of the effective attraction between particles at short distances. Here it provides a mechanism to borrow energy. According to the uncertainty principle in this form, energy of any amount can be borrowed from Nature, but only for a time interval inversely proportional to the amount borrowed. The more you borrow, the sooner you have to return it.

If an inter-particle potential acts like a long-term bank account, as we described earlier in this chapter, then the quantum uncertainty principle acts like a short-term bank account. Neither of these accounts charges interest or an administration fee.

With the uncertainty principle, a no-particle state cannot remain so for very long, because energy can and will be borrowed to create particles. It is true that these created particles do not last very long; they must disappear when the energy is returned to Nature after a short time. This action leaves behind a plethora of bubbling particles called *virtual particles*, coming and going, appearing and disappearing. The virtual particles interact among themselves and that alters the energy of the initial state. Therefore, there is no such thing as a no-particle steady state, and a steady state with the lowest energy is not a state with absolutely no particles in it.

Vacuum energy density is the energy per unit volume in the vacuum state. We do not know whether our vacuum has a non-zero energy density or not, but if it has, it would be subject to gravitational interaction like any other energy. The possibility of having a vacuum energy was first conceived by Albert Einstein in 1917, but he called it a *cosmological constant*. A cosmological constant is simply equal to some known positive number times the energy density of the vacuum. Here is why he introduced it.

We remember from Chap. 6 that a gravitational force is always present between two massive particles. With the equivalence of mass and energy through the relation $E = mc^2$, this force should also exist between two chunks of energy. Thus the photon, which is massless but carries energy, must still experience a gravitational attraction by the sun when it passes close to its surface. Einstein suggested using this effect to verify his general theory of relativity. This experiment was subsequently carried out by Arthur Eddington and his team, and confirmed that Einstein's prediction was correct.

With that idea proven, a positive energy density in the vacuum must also experience a gravitational force. Unlike the usual gravitational force, this one acts to repel, *as if* it were anti-gravity (see Chap. 15 for an explanation of this bizarre behavior). In 1917, when Einstein introduced the cosmological constant, Hubble had not yet discovered the expansion of the universe, and it was commonly believed at that time that the universe was static and eternal. However, due to the mutual gravitational attraction of galaxies which draws them towards one another, the universe must shrink and cannot remain static, unless there is a repulsive force present to counteract gravity. That is why Einstein introduced this positive vacuum density, which as remarked above acts repulsively.

Einstein was a bit uneasy at that time to bring in this new concept without additional supporting evidence. This is shown in his letter to Paul Ehrenfest early that year, which said, "I have again perpetrated something relating to the theory of gravitation that might endanger me of being committed to a madhouse." After Hubble's discovery, he regretted the introduction of the cosmological constant even more, and was supposed to have said "that was the biggest blunder in my life." However, I understand that this particular statement cannot be found in his published work and letters, so it is unclear whether he had really made such a strong statement himself or not.

In any case, he really did not have to be sorry. Although the universe is no longer static as he believed, the cosmological constant may just have been resurrected and renamed dark energy. The difference is that in this reincarnation, there is more dark energy than needed to balance the attractive gravitational force, so the universe ends up accelerating.

Whether dark energy density is really a cosmological constant or not depends on whether it is time-independent. A vacuum is the same at all times, so if the dark energy density is a vacuum energy density it must stay constant. Present observation shows that it is consistent for the dark energy to be a cosmological constant, but only refined observations of the future will be able to tell for sure.

Both Pauli and Einstein were quite hesitant to postulate the existence of new objects out of the blue — anti-neutrinos for

Pauli and the cosmological constant for Einstein — but both were eventually proven to be correct or useful. This shows how attitudes in doing physics have changed in less than a century. If you open a journal of physics nowadays, especially in the fields of particle physics and cosmology, you will see that people have no hesitation whatsoever in postulating new objects and new theories, sometimes with very little physical evidence or even a convincing motivation.

Let us now get back to dark energy and discuss some of its puzzling aspects.

The present universe, according to the present best estimate, consists of 73% dark energy, 22.5% dark matter, and 4.5% ordinary matter (Chap. 10). Photon and neutrino energies are negligible. If dark energy is a vacuum energy density which is independent of time, it must be tiny compared to neutrino and photon energy densities in the early universe because the latter vary with the time-dependent scale factor a like $1/a^4$ (Chap. 13). Yet this tiny amount of dark energy was just enough to make the universe almost flat today. How could the early universe be so smart and do that? And, why do we happen to live at a time when dark energy density is dominant?

In the same vein, one might also ask why we live at the present time when the photon energy is negligible, and not at an earlier time when photon energy dominated over matter and dark energies (Chaps. 13 and 15)? We know the answer to this one: the universe was so hot at that time that not even atoms could be formed, much less life. However, there is no similar argument for dark energy that I am aware of, so the corresponding question for dark energy is very difficult to answer.

Moreover, it is impossible for the present theory of particle physics (see Chap. 17) to obtain the right amount of vacuum energy density consistent with cosmological observations. This difficulty is commonly known as the *cosmological constant problem*. We do not know why this is so, but it may just be an indication that we do not know how to treat quantum gravity correctly.

13

Heat and Temperature

Heat consists of the *random* kinetic energy contained in systems containing a large number of particles.

Take a container of molecules, or particles. Although the container may be stationary, the particles inside are in constant motion, colliding with one another, and with the walls of the container. As a result, the direction of motion of each particle changes all the time, making its motion random in direction.

Collisions can transfer energy from one particle to another. Over time, whatever energy there is in the container will be equally shared among all the particles. The system is said to have reached a *thermal equilibrium* when that happens.

The *absolute temperature* of a system is just a measure of the average amount of kinetic energy carried by a particle. Since this is always positive, it must differ from the usual temperature measured on either the Fahrenheit or the Celsius scales because both of them can go negative. Actually, the gradation of the absolute temperature is designed to be equal to the gradation on the Celsius scale; it is equal to the Celsius temperature plus 273.15 degrees. This absolute temperature scale is known as the *Kelvin scale*, usually abbreviated by the letter K, just as the Celsius scale is abbreviated by the letter C. Thus, the freezing point of water, at 0 degrees C, is 273.15 K. A hot day of 27 degrees C is about 300 degrees K. This also means that no matter how cold it is, the temperature can never get below –273.15°C. From now on, when we talk about temperature, we will always mean the absolute temperature and always use the Kelvin scale, since the laws of thermal physics are naturally expressed on this temperature scale.

The average kinetic energy of a monatomic molecule is equal to $3kT/2$, where T is the absolute temperature, and k is called the Boltzmann constant, whose magnitude is 1 eV/1.16 × 10^4 K.

Instead of the Kelvin scale, we often express temperature on an energy scale. If we say the temperature is so many eV, what we mean is that kT is so many eV. From the magnitude of k above, we see that the conversion is: 1 eV corresponds to 1.16 × 10^4 degrees Kelvin. That is almost twelve thousand degrees, roughly twice the surface temperature of the sun, so 1 eV of temperature is a fairly high temperature on the human scale.

The constant collisions of particles in the container will create new particles if sufficient thermal energy is available and no conservation law is broken. For example, with the rest mass of an electron being 0.51 MeV (actually mc^2), electron–positron pairs will be created when the temperature is above 1 MeV or so.[1]

Since photons are massless, they can be created at any temperature. At high temperature more energy is available so more photons can be created. Calculation[2] shows that its *number density* (number per unit volume) is proportional to the third power of the temperature (T^3), and its *energy density* (energy per unit volume) is proportional to T^4. That also means that the average energy per photon is proportional to T.

Particles with a mass m can be pair created when the temperature is above $2mc^2$ or so. For a temperature way above that, the particles are relativistic and their rest mass becomes negligible.

Like the photons, their number density is again proportional to T^3 and their energy density is proportional to T^4.

At a given temperature, the *average* kinetic energy of *each particle* is fixed, but since energy is changed by constant collisions, at any given time its kinetic energy may be smaller or larger than the average. The distribution of energy at any given temperature can be calculated by statistical mechanics. It depends on the mass of the particle, as well as the temperature. It is different for fermions (matter particles such as electrons and nucleons) than for bosons (such as photons).

The distribution for photons, known as the black-body distribution or the *black-body spectrum,* was the first distribution to be discovered. Its calculation by Max Planck in 1900 was actually the event that launched the quantum era.

Remember from the beginning of Chap. 12 that the energy of a photon is inversely proportional to its wavelength. Thus, its energy distribution can be translated into a wavelength distribution; its average energy at a given temperature can be translated into an average wavelength. For visible light, different wavelengths give rise to different colors of light, with violet at short wavelengths and red at long wavelengths.

An example of the relation between temperature and wavelength can be seen in daily life. If you turn on your electric stove at home, at first you feel the heat but you do not see any color. This is when you get the invisible, long wavelength infrared radiation. As its temperature rises, it begins to glow red, which has a shorter wavelength than the infrared. At a higher temperature still, which your electric stove cannot reach but you might see it in the blast furnace of a steel mill, or the surface of the sun, the color turns orange and yellowish, signifying still shorter wavelengths of light being emitted.

Stars in the sky come in different colors, some reddish and others bluish. These different colors reveal different surface temperatures of the star, with the bluish ones much hotter than the reddish ones. Our sun emits sort of orange color, corresponding to its surface temperature of about 6,000 degrees.

The blackbody distribution of photons is shown in Fig. 34. The x-axis represents wavelength in 10^{-9} meters, or nanometers (nm). The y-axis is proportional to the energy density per unit wavelength, so that the area under a curve is proportional to the energy density from all wavelengths, i.e. the brightness. Note that the peaks shift to shorter wavelengths or higher energy as the temperature rises, in agreement with the fact that the average energy of a photon is proportional to its temperature. Note also that the area under each curve grows very quickly with rising temperature, reflecting the fact that its energy density is proportional to the fourth power of temperature.

Since photons move really fast, they bounce hard on the walls, exerting a large *pressure* on them. Calculation shows that the

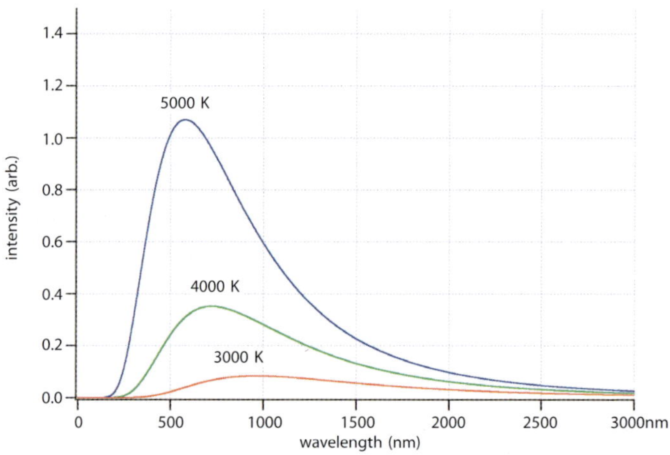

Figure 34: Blackbody radiation curves giving intensity distribution per wavelength at different temperatures.

pressure is equal to one third of its total energy density. The same is true for a gas of relativistic particles.

The pressure exerted by a gas of slow-moving, non-relativistic particles is negligible in comparison, because they bounce so softly on the walls. As an approximation, we usually take the pressure of non-relativistic particles to be zero in cosmology.

Lastly, I want to mention one very important result for photons and relativistic particles. If you slowly vary the volume of the container, making sure that no heat enters or leaves it in the process, then the same amount of thermal energy is shared in a different volume so the temperature must change. It turns out that the resulting temperature T is inversely proportional to the linear size of the volume.[3]

At first sight this result may be a little disconcerting. Since energy density is proportional to T^4, the total energy in a container of volume V is proportional to T^4V. According to what we have just said, T^3 is proportional to $1/V$, so the total energy is proportional to T. This means that energy in the volume is lost when the temperature is lowered as the volume expands. Where has it gone to?

The answer comes when one realizes that relativistic particles exert a substantial pressure on the walls of the container: the energy lost is being used by the pressure to do work to expand the volume.

When you apply this result to the universe,[4] whose linear size is proportional to the scale factor a introduced in Chap. 9, one concludes *that the temperature T of the universe at any time is proportional to* $1/a$.

Remember that the total energy of the photons in the universe is proportional to T, and thus proportional to $1/a$. Since the energy of a photon is also inversely proportional to the wavelength, it follows that the wavelength of the photon is proportional to a.

This is simply a reflection of the fact that as the universe expands, space is stretched so the wavelength of the photon is stretched by the same amount as well. We can also turn this argument around to understand why the total energy of the photons decreases with T.

14

The Noisy Universe

The fundamental knowledge of physics explained in the last three chapters will now be used to explore the universe in more detail. To begin with, let me relate the second important landmark of cosmology: the discovery of *cosmic microwave background radiation*, or CMB for short.

In 1963, Arno Penzias and Robert Woodrow Wilson, two radio astronomers working at the Bell Laboratories in Holmdel, New Jersey, USA, decided to use the 20-foot horn-reflector antenna, originally built to receive signals bounced off the Echo satellite, to study astronomy. They found an annoying noise which they could not get rid of no matter how hard they tried, including cleaning off the pigeon droppings left on the antenna and getting rid of the two pigeons. This unexpected electromagnetic noise turned out to be cosmic in origin, left behind by the reverberations from the original Big Bang. This discovery is so important that they were awarded the Nobel Prize for Physics in 1978.

The original and the subsequent measurements reveal that the wavelength distribution of this electromagnetic noise obeys a black-body spectrum with a temperature of 2.725 K. It is the same in all directions, to an accuracy of several parts in a thousand. The dominant wavelengths at this temperature are in the microwave

Figure 35: Penzias (right) and Wilson (left) in front of the antenna they used to discovered the CMB.

and the far infrared region, which are partly absorbed by the earth's atmosphere. To improve accuracy of the data, one needs to find a locale where atmospheric effects are minimized, in the arctic region or above the earth's atmosphere. This is best done with satellites, where the accuracy reaches parts in a million at the present.

The first dedicated satellite, COBE (COsmic Background Explorer), was launched in 1989. At the time of writing in 2007, the satellite up there is the WMAP (Wilkinson Microwave Anisotropy Probe), launched in 2001. A third satellite, Planck, is scheduled to be launched some time in 2008.

Already with an accuracy of one part in a thousand, obtained by flying a U2 spy plane high above the atmosphere in 1976 and 1977, one began to see a slightly different temperature in two opposite directions in the sky. Half the sky is on average slightly warmer than the other half, an effect due to the motion of our sun in the universe. In the direction towards which the sun is moving, light waves are compressed, their wavelengths shortened, and that corresponds to a higher temperature. In the opposite direction, waves are stretched and wavelengths lengthened, corresponding to

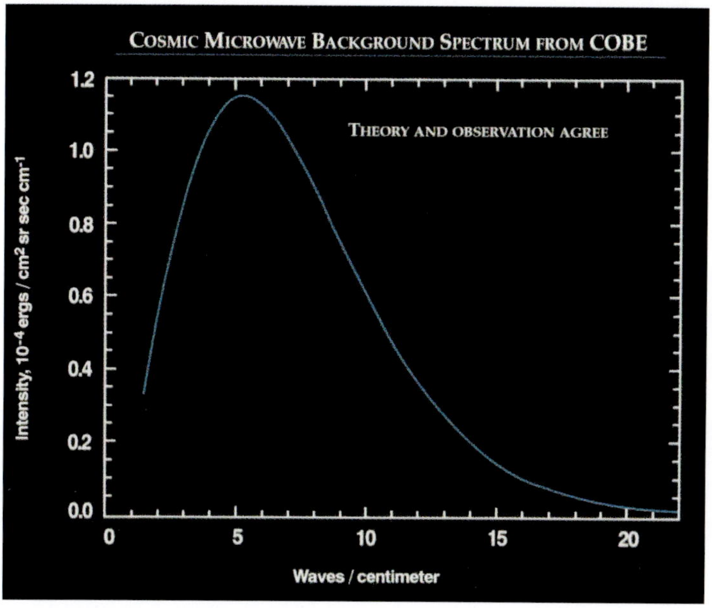

Figure 36: The perfect blackbody spectrum of the CMB taken from the COBE satellite,[1] corresponding to 2.725 degrees Kelvin. The error bars are smaller than the width of the line.

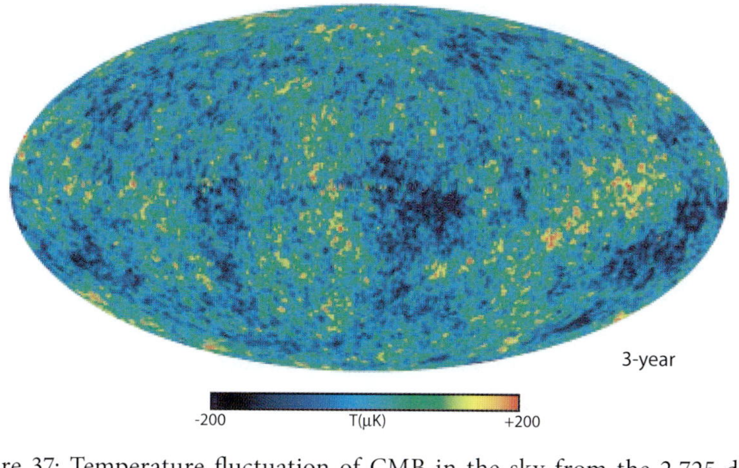

Figure 37: Temperature fluctuation of CMB in the sky from the 2.725 degree background, three-year data taken by the WMAP satellite.[2] Blue regions have slightly higher temperature than red regions.

a lower temperature. This kind of bipolar distribution is called a *dipole*.

With an accuracy of parts per million, obtained by the COBE and WMAP satellites, the anisotropy (differences in different directions) becomes much more complicated, but that complicated pattern can be interpreted as an acoustic oscillation set up by quantum fluctuations at the beginning of the universe. This complicated but small fluctuation turns out to be very important, because much information about the universe can be extracted from it, as will be discussed in Chap. 17.

I will discuss now how the cosmic microwave background radiation comes about. A more detailed discussion, especially about the fluctuation seen in Fig. 37, can be found in Chap. 17.

Just after the Big Bang, the universe was very small, but very hot. Atoms and complex nuclei could not exist because they were all torn apart by the intense heat into nucleons and electrons.

This equal mixture of positively charged and negatively charged particles is called a *plasma*.

Since photons can easily be absorbed and emitted by charged particles, in the dense environment of an early universe, no photon could last very long or travel very far before being reabsorbed. Although the universe was very bright at that time, it was also very opaque because the photons reaching your eyes must come from only a very short distance away.

This opaqueness is present even at the much lower temperature of the sun. The photons reaching us all come from a surface layer of the sun, known as the photosphere. The 6,000 degree temperature of the sun quoted before is only the temperature at its surface. The temperature at the center of the sun is some 15 million degrees, but because of the opaqueness we cannot see it directly.

About 400,000 years after the Big Bang when the temperature of the universe was reduced to approximately a quarter of an electronvolt, corresponding to a scale factor of about a thousandth and a redshift of about 1,000, the universe finally became sufficiently cool for electrons to be captured by the nuclei to form neutral atoms. This allowed the trapped photons to escape, which come to us today as the observed CMB.

The time when neutral atoms were formed and photons finally came out unimpeded is known either as the time of *recombination* (to form neutral atoms), or *decoupling* (of the photon from matter). We will come back in Chap. 17 to discuss what happened at that time and the wealth of information about the universe the CMB gives us.

15

A Short History of the Universe

The history of civilization is passed down from generation to generation, recorded in documents and cross checked with archaeological evidence whenever possible.

The history of the universe must be deduced from observational evidence and the laws of physics, because nobody was around to tell us what happened in that distant past.

Archaeology on earth relies on geological layers and dating techniques to determine the age of an ancient artefact. Thanks to the finite speed of light, digging deep is equivalent to looking far in astro-archaeology. Large telescopes are required to detect the dim lights from far away, but no matter how large the telescope is, we can never look past the age of decoupling into the opaque plasma. Other means must be found to probe events that are even older, and some of these techniques will be discussed in Chaps. 17 and 19.

In this chapter I will outline a short history of the expanding universe, beginning at the time after the Big Bang when the universe was very small and very hot.

This history can be divided into three periods: ancient history, known as the '*radiation-dominated era*'; medieval history, known as the '*matter-dominated era*'; and contemporary history, which is the '*dark-energy dominated era*.'

Before all these there is a pre-historic period known as the '*inflationary era*.' It is during this pre-historic period that the nearly empty universe grew to have all the energy of the present universe. This era is extremely short so its time span will be completely ignored in the rest of this chapter. Inflation will be discussed in the next chapter.

In human history, time is usually measured backwards from today, or from the birth of Christ. In astronomy, time is specified in years since the Big Bang, but it can also be specified by the redshift z, the scale factor a, or the temperature T of the universe at that moment. As mentioned before, the present age is $t_0 = 13.7$ billion years after the Big Bang. It has a temperature of $T_0 = 2.725$ K, a redshift of $z_0 = 0$ and a scale factor of $a_0 = 1$. At any other time t, these three quantities are related by $T = T_0/a = T_0(z + 1)$. The redshift is frequently used because it is directly measurable (Chap. 9), and the temperature is most useful in describing the appearance of certain physical events. The relation between the scale factor a and time t depends on the energy contents of the universe and will be discussed later in this chapter.

Let us start with the ancient history, when the universe was very small and hence very hot. The high temperature caused all atoms and nuclei to be torn apart, and most particles to be relativistic like the photon. Relativistic particles exerted a large pressure on the environment (Chap. 13), and generally behaved like a photon in many other aspects. Hence, this era characterized by the dominance of relativistic particles is said to be radiation dominated. It is during this era that an excessive amount of matter materialized over anti-matter (Chap. 18).

The energy density of relativistic particles was proportional to T^4 (Chap. 13), or $1/a^4$, because the size of the universe measured by the scale factor a was *inversely* proportional to T. Hence, the total energy, which was proportional to T^4a^3, became proportional to T. It decreased as the universe expanded. As explained in Chap. 13, the energy was not lost, but was used to overcome the pressure in the expansion.

This era was very short. It lasted less than 60,000 years, until a redshift of $z_{eq} = 3233$, which corresponded to a temperature of $T_{eq} = 0.74$ eV and a scale factor of $a_{eq} = 3.1 \times 10^{-4}$, 0.03% of the present size.

As the temperature dropped, many particles gradually became non-relativistic and ceased to exert pressure on their surroundings. Medieval history formally began at $z_{eq} = 3233$ when the energy density of non-relativistic matter became equal to that of the relativistic particles consisting of photons and neutrinos. Beyond that, the era became matter dominated, with the energy density dominated by the rest mass of the nucleons.

During this period, energy was dominated by non-relativistic matter, each of whose particles had a total energy very close to its rest energy. Neither the total number of non-relativistic particles nor their individual energy could change with time; hence, the total energy of the universe remained fixed in this period. No energy was lost as in the previous period because there was now no radiation pressure to overcome. Since the total energy was fixed and the volume varied like a^3, which was proportional to $1/T^3$, the energy *density* in this period was proportional to T^3, or $1/a^3$.

The matter-dominated era was by far the longest of the three periods. The universe spent almost its entire time in this era. It was during this era that decoupling occurred (Chap. 14), about 400,000 years after the Big Bang, at a redshift of $z_* = 1089$. That corresponded to a temperature of $T_* = 0.256$ eV and a scale factor of $a_* = 9.2 \times 10^{-4}$. After decoupling, light streamed freely into space forming the cosmic background radiation (CMB), we see today. It was during this period that galaxies and stars began to form; even the earth was formed in this period.

As mentioned in the last chapter and shown in Fig. 37, there is a small (parts per hundred thousand) fluctuation in the CMB, which is believed to have come from quantum fluctuations in the inflationary era (Chap. 17). This fluctuation produced a slight overdensity at some locations, and a slight underdensity at others. Since gravitational attraction grew with density, surrounding matter and energy tended to be drawn into the overdense regions. The stuff so gathered also tended to settle down to the center, further increasing the density and gravitational binding of the region. The gravitational potential energy thus released from the collapse was converted into thermal energy, increasing the temperature at the center. Eventually, these regions would grow to be so large, so dense, and so hot that galaxies and stars could be formed.

However, this process could not begin before decoupling, because light trapped in the charged plasma exerted a pressure to prevent such a collapse from taking place. Instead, the gravitational compression counteracted by the photonic pressure set up an acoustic oscillation which lasted until decoupling (Chap. 17). The image of the acoustic oscillation caught at the moment when light became free is what we see in Fig. 37.

After decoupling, pressure was released because photons were no longer trapped; then gravitational collapse could commence to form galaxies, stars, and planets.

It is truly amazing that we can trace the seed of galactic formation all the way back to the inflationary era, the prehistoric period which we will discuss in the next chapter. Galaxy formation does not occur until a redshift of 20 or so, corresponding to a temperature and an energy scale of about 5 milli-electronvolts. The energy scale in the inflationary era is probably as high as 10^{16} GeV, a full 28 orders of magnitude larger, yet whatever fluctuation originated at this high energy scale is supposed to have lasted and become the seed of galaxy formation so much later!

Figure 38 shows a computer simulation of galaxy formation by the Virgo Consortium, using cosmological parameters measured in CMB (Chap. 17) and other observations. Bright spots represent overdense areas and dark spots underdense regions. We can see from this simulation the formation of filaments, and the emergence of a galaxy at their intersection. The result compares favourably with the data from galaxy surveys mentioned below.

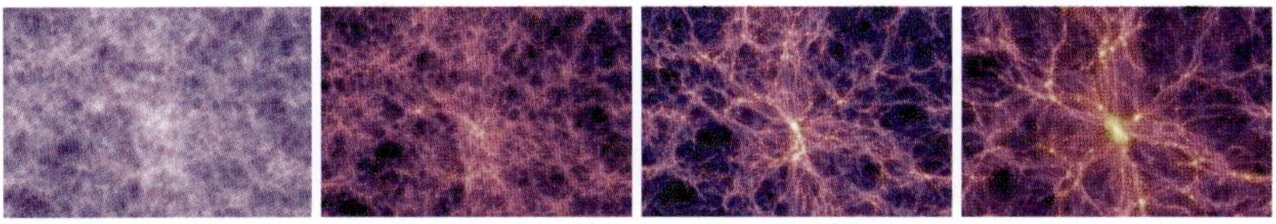

Figure 38: A computer simulation of galaxy formation by the Virgo Consortium. From left to right, the four panels show the evolution respectively at redshifts z = 18.3, 5.7, 1.4, and 0.0.

It is not yet certain when galaxies began to emerge, but when they did, the ultraviolet light from the first massive stars would reionize their surroundings, making the universe once again partially opaque to the CMB. The time at which that occurred affects the CMB pattern. Analyzing it, the present estimate puts the reionization at about a redshift of $z_r = 20$, with large error bars.

Our knowledge of galaxies is being vastly improved by recent systematic mappings and surveys. One of these is the Sloan Digital Sky Survey (SDSS), using the 2.5 meter Sloan telescope at the Apache Point Observatory in New Mexico. The other is the

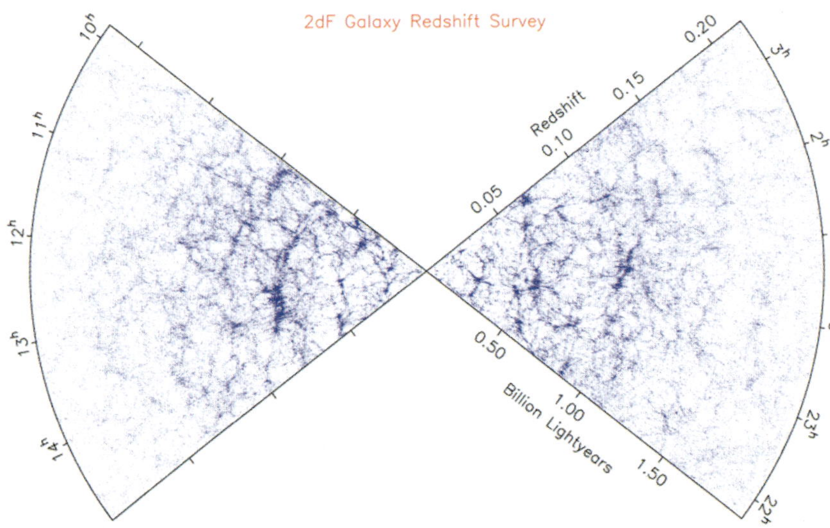

Figure 39: More than 200,000 objects shown in the 2dF Galaxy Redshift Survey.

2 degree Field Galaxy Redshift Survey (2dFGRS), using an Anglo-Australian telescope in Australia. Figure 39 shows the result of this survey of more than 200,000 objects, out to a redshift of more than 0.20.

Contemporary history began at $z_{de} = 0.39$, *when* the dark energy amount became equal to that of the matter energy. As mentioned before, dark energy presently comprises 73% of all energy densities of the universe, with the remaining 27% residing in ordinary and dark matter. Photon energy is negligible and neutrino energy is small. Dark energy is characterized by a nearly constant energy density,[1] which changes little with time. In contrast, the energy density of radiation is proportional to $(z + 1)^4$ and the energy density of matter is proportional to $(z + 1)^3$, so both became larger in the past than at the present, and the further back in history we go, the larger the redshift z is, and the larger they become. Clearly at some point their sum would be equal to the dark energy density, although they are smaller at the present. Given the present proportions of the various constituents in the universe, it is not difficult to estimate that this occurs at a redshift of about 0.39.[2]

Dark energy causes the universe to accelerate at the present time.[1] In fact, the acceleration already began towards the end of the matter-dominated era.

In the rest of this chapter we will discuss how fast the universe expands, what determines its expansion rate at any given time, and what fixes the curvature of the universe.

If the universe had no gravity, no pressure, and no dark energy, then it would always expand at a speed of 72 km/s/Mpc, as determined by the Hubble Law (Chap. 7). In that case the universe would always grow at the same rate and the scale factor a would be proportional to time t.

Gravitational attraction and the pressure exerted by other galaxies both work to retard the expansion. As a result, the universe cannot expand as fast as t in the radiation- and matter-dominated eras. As will be explained below, the scale factor a is proportional to $t^{1/2}$ during the radiation era and $t^{2/3}$ during the matter era. Note that 2/3 is between 1/2 and 1, because the pressure is absent in the matter era to retard the expansion.

Dark energy is characterized by a negative pressure.[1] If dark energy is the vacuum energy (Chap. 12), then the negative pressure is equal to its energy density in magnitude. While a positive pressure in the radiation era retards the expansion, a negative pressure does the opposite and makes the universe expand faster. As the dark energy density becomes more important deeper into the contemporary period, the acceleration increases with time. When the dark energy completely dominates, a attains an exponential growth in time, as we shall see.

Negative pressure is a very strange thing, something that we have no daily experience of! We are familiar with the positive pressure shown in Fig. 20 — look how hard the little girl has to blow to inflate the balloon in order to overcome the positive atmospheric pressure surrounding it. If the atmospheric pressure were negative, then not only would the surrounding atmosphere automatically suck out the balloon to inflate it, but the inflation would become faster and faster just like the present universe and the universe during the inflationary era (Chap. 16).

To see how these expansion rates are arrived at, we consider the energy of an object in the universe. This object is carried outward by the expansion of the universe, so it has a kinetic energy. It is being attracted gravitationally by all the galaxies, so it has a potential energy. The sum of these two is conserved, from which we can derive an equation known as the *Friedmann equation*,[3] governing the evolution of the universe. This is the most important equation in cosmology, and it determines the expansion rate of the universe.

If we let da/dt denote the rate of change of the scale factor a, then the Friedmann equation is

$$\left(\frac{da}{dt}\right)^2 - \frac{8\pi G\rho a^2}{3c^2} = -k$$

where ρ is the energy density of the universe, G is the Newtonian gravitational constant (Chap. 6), and $-k$ is a constant related to the conserved total energy of that chunk of the universe. The first term on the left is proportional to the kinetic energy of the small chunk, and the second term is proportional to its gravitational potential energy. For that reason I shall refer to these two terms respectively as the 'kinetic' and the 'potential' terms.

If k is positive, then the potential energy of every chunk in the universe is larger than its kinetic energy, so that must also be the case for the universe as a whole. The universe is then said to be *closed.* The spatial curvature of the universe in that case can be shown to be positive, like that of a sphere.

If k is negative, then the universe has more kinetic energy than potential energy, and it is said to be *open.* The spatial curvature of the universe is then negative, like that of the middle of a saddle.

If $k = 0$, then the potential energy is exactly equal to the kinetic energy. The universe is said to be *critical,* and the spatial curvature in that case is zero. Namely, the universe is flat.

The fractional expansion rate $(da/dt)/a = H$ is called the *Hubble parameter.* This parameter varies with time because both a and da/dt do. Its value at the present time is simply the Hubble constant $H_0 = 72$ km/s/Mpc discussed in Chap. 7.

For a critical or flat universe, $k = 0$, the Friedmann equation becomes $H^2 = 8\pi G\rho/3c^2$. The energy density ρ_{crit} computed from this equation by setting $H = H_0$ is called the *critical density.* If the density of the universe at the present time is larger than ρ_{crit}, then $k > 0$ and the universe is closed. If it is less than ρ_{crit}, then $k < 0$ and the universe is open. If it is equal to ρ_{crit}, then $k = 0$ and the universe is critical and flat.

As mentioned before, our universe is flat within observational error, with 73% as dark energy, 4.5% as ordinary matter, and 22.5% as dark matter. So $k = 0$ and our universe is critical, at least within observational errors.

During the radiation era, ρ is proportional to $1/a^4$. Substituting this back into the Friedmann equation, we see that $a(da/dt)$ must be a constant. Using calculus, we conclude from this that a is proportional to $t^{1/2}$.

During the matter era of a flat universe, ρ is proportional to $1/a^3$. Substituting this back into the Friedmann equation,[4] we see that $a(da/dt)^2$ must be a constant; hence, a is proportional to $t^{2/3}$. From this it follows that if dark energy is neglected, the age of the universe is given by 2/3 of the inverse Hubble constant $1/H_0$.[4]

Similarly, we can work out the time dependence of a in the contemporary period, when dark energy dominates.[5]

Finally, during the inflationary era when ρ is a constant, a grows exponentially in time.[6]

Before the contemporary era, dark energy was completely negligible. If we ignore dark energy, then the fate of the universe is similar to the fate of a cannon shell shot upward from a big gun. If its initial velocity is so big to render its kinetic energy larger than its gravitational potential energy, it will escape the earth's gravitational confine and fly into outer space, like what happens

Figure 40: A 16-inch gun with a smooth bore, lengthened from its standard 20-meter length to 37 meters. It was built by Gerald Bull in the 1960s at Barbados to do high altitude research.

to the small chunk of an open universe. In principle a gun can be built to launch a vehicle into space that way, but the one in Fig. 40, one of the biggest guns at that time, was not large enough to do so. If the kinetic energy of the shell is smaller than its potential energy, the shell will simply fall back to earth after reaching its maximum height, like the dust particle in a closed universe. In between, when the kinetic and potential energies are equal, it will make it to infinity, but barely, just like the dust particle in a critical universe. The initial velocity in this case is known as the *escape velocity*. It is an analogy of the *critical density* of the universe.

You might have seen the Yin-Yang symbol in Fig. 41 before I claim it can be used to represent the Friedmann equation.

This symbol depicts the primordial division of the whole into a dichotomy, or a duality, and conversely the unification of the dichotomy into the whole. According to that, the universe started out as an amorphous body known as the 'Tai Ji' ('Tai Chi'), symbolized by the circle. The dichotomy is known as 'Liang Yi' (two shapes or two phases), signifying the dynamical balance of the universe. The most popular interpretation of the 'Liang Yi' is 'Yin' (the shady) and 'Yang' (the bright), but scholars have pointed out that there were at least six other interpretations, including heaven and earth, soft and hard, odd and even, etc. All these are supposed to have something to do with 'I Jing' ('I Ching', the Book of Change), the most ancient of Chinese classics which describes the order of the universe out of seemingly random events. The 64 possible outcomes of the universe and a person's fate and fortune are labeled by a pair of eight primitive symbols known as the 'Ba Gua.'

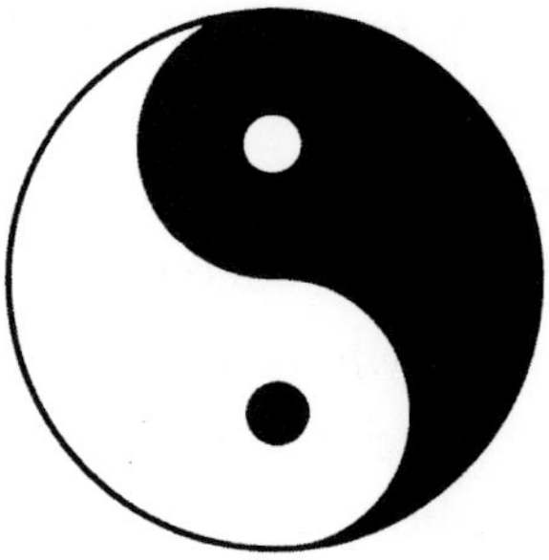

Figure 41: The symbol of Tai Chi and Yin Yang.

It is said that this connection of Tai Chi and Liang Yi to Ba Gua came from Confucius. He reaches one from the other by a binary multiplication: starting from 1 (the Tai Chi), it becomes 2 (Liang Yi), then 4 ('Si Xiang', the four directions), and finally 8 (Ba Gua).

The Tai Chi symbol itself has nothing to do with Buddhism, but it is often identified with Taoism. When Buddhism first came to China, because of similarities between certain concepts in Taoism and Buddhism, the more familiar concepts of Taoism were often borrowed to translate the foreign concepts of Buddhism based on Sanskrit and Indian culture. Partly because of that, there is often a curious mixture of Buddhism, Taoism, and Confucianism in China.

Tai Chi was primordial and amorphous, but not empty enough to explain the emptiness in Buddhism. Partly to accommodate that deficiency, the Chinese author Dun-Yi Zhou (周敦頤) (1017–1073 CE) invented a somewhat equivalent concept of 'Wu Ji,' meaning roughly limitless or eternal, and advocated that Tai Ji came from Wu Ji.

With modern cosmology I can provide another interpretation of this highly philosophical complex. The universe just before inflation is in the state of Tai Chi; the presently unknown state of the universe preceding inflation is in the state of Wu Ji. The dichotomy in Liang Yi stands for the negative potential energy and the positive kinetic energy, so all the subsequent developments of the universe, from pre-history to contemporary history, are simply different divisions of the whole into these two, with the total added up to be the same original Tai Chi. With this interpretation, the Tai Chi symbol simply represents the Friedmann equation; no wonder it can describe the change and the order of the universe as I Ching claims! Oh, I suppose I should also attribute the different outcomes based on Ba Gua to different initial conditions of the Friedmann equation.

16

Inflation

The scenario described in the last chapter is known as the *Classical Big Bang Theory*. It is a very successful theory capable of explaining many different phenomena in the universe, but as pointed out by Alan Guth in 1980, it also has three rather serious problems: the *flatness problem*, the *horizon problem*, and the *monopole problem*. I shall describe the first two but not the third, because it will take us too far afield to explain what a monopole is. The *theory of inflation* proposed by Guth was aimed at solving these problems.

The Flatness Problem

Why is the universe spatially flat? Of all the k values in the Friedmann equation, why does it choose the critical value $k = 0$? This is the flatness problem.

When Guth raised this question in 1980, dark energy had not yet been discovered, so at that time the universe actually appeared to be open, with only 27% or so of the critical density. Even so there was still a flatness problem in the early universe for the following reason.

Since gravity and pressure slow down the expansion of the universe, the 'kinetic' term of the Friedmann equation is largest at the earliest moment. In fact, since $a(da/dt)$ is a constant in the radiation era, da/dt becomes infinite when a goes to zero. Whenever the kinetic term is large the potential term has to be large as well because their difference is a constant $-k$.

In the early universe when a is small, each of these two terms is so much bigger than $|k|$ that it does not matter then whether we set $k = 0$ or not. In other words, although the universe appears to be open at this moment, if we go back sufficiently far in time, it is indistinguishable from a flat universe. Why the universe chose to be so flat at the beginning is then the flatness problem.

If you imagine the universe to look like the balloon in Fig. 20, then a flat surface corresponds to a large balloon, so perhaps another way of asking this question is why our universe is so large.

The question asked this way is more intuitive, but it is also a bit vague. Large compared to what? In the case of Fig. 20, one can say that the balloon is large if it is much bigger than the little girl there, but for the universe, what should we compare its size to?

Arguments have been advanced that one should compare it with the *Planck length*, which is about 10^{-35} meters. That might sound ridiculous, because not only is the universe vastly larger than that — even a tiny nucleus (the size of a nucleus is about 10^{-15} m) is — but why would one compare anything with this tiny length at all?

The argument runs as follows. There are three fundamental constants in physics, associated with the three important discoveries in the first quarter of the 20th century: Planck's constant $\hbar$ in quantum mechanics (which defines the scale when quantum effects set in), the speed of light c (which determines when special relativity is needed), and the Newtonian gravitational constant G (which is the basic constant used in general relativity). Using algebraic combinations of these three fundamental constants $\hbar$, c, G, one can construct three natural units of time, length, and mass, known respectively as the Planck time, the Planck length, and the Planck mass.[1] In particular, the Planck length is about 10^{-35} meters. To continue the argument, the reason why the Planck length is relevant in cosmology is because gravity is the main driving force, whereas in other branches of physics, for example, nuclear physics that determines the size of a nucleus, G is not involved but some other constant such as the mass of a nucleon or some other particle, and from these other constants a natural length scale such as 10^{-15} m could be arrived at. These formal arguments of course do not address the dilemma of how a nucleon could possibly be larger than the universe if the size of the universe is indeed the Planck length.

This argument as it stands assumes quantum mechanics to be relevant in cosmology. But this is not so because the Friedmann equation of the last chapter contains G and c but not $\hbar$. Without $\hbar$, one cannot arrive at the Planck length, so that tiny number cannot be the thing to compare with when one wants to know why the universe is so large. For that reason it is better to ask why the universe is so flat, rather not so large. We can use the dimensionless number k in the Friedmann equation as a measure of the curvature.

On the other hand, the argument involving the Planck length is not a complete nonsense either, because near the Big Bang the universe was so small that quantum effects had to be taken into account. At that early point the relevant length was the Planck length, and one can then ask how a universe of that size can grow out of it to become our present universe which is so huge. We do not know how to answer that question because at that size quantum gravity, which we are so ignorant of, becomes important. However, if we assume somehow that the universe can grow from the size of the Planck length to something a thousand times bigger or so, then the theory of inflation, which solves the flatness problem, will also tell us how the universe grows from there into the present gigantic size.

The Horizon Problem

I have already mentioned that at the beginning of the universe, the expansion rate da/dt is proportional to $1/a$, which can grow indefinitely large as a approaches zero. In particular, at the beginning the universe, it is so large that the universe can expand faster than the speed of light.

This is not a violation of special relativity, which requires that no object or signal be allowed to travel faster than the speed of light. Expansion of the universe should be thought of as a stretching of the underlying space and not the speed of any object, nor can it be used to propagate a signal faster than the speed of light.

However, as space is constantly "running away," it is hard for a light signal to catch up, so any signal sent at the beginning of the universe can only reach so far after a finite amount of time. In fact, calculation shows that a signal emitted at the Big Bang would have traveled only about a couple of degrees across the sky by the time of decoupling. That means that the CMB coming from two directions of the sky separated by much more than a couple of degrees are not causally related, so there is no reason for them to have the same strength and the same frequency, which they do as we learned from Chap. 14.

This is the horizon problem, so called because the horizon is the farthest point light and any causal influence can go to or come from.

One way out of this difficulty is to have the universe already homogeneous at the very beginning of the classical Big Bang.

Then there is no need to rely on the propagation of a dynamical influence to smear out the initial inhomogeneities. But the classical universe started out with a big explosion, and it is hard to imagine how an explosive event can be homogeneous.

To solve these problems, Guth proposed the *theory of inflation*. It provides a mechanism for the universe to grow exponentially to a size 10^{25} times or more during a very short time period. Since the size grows exponentially, the expansion velocity also grows exponentially by about the same factor. This causes all the galaxies to rush out so fast at the end of inflation that it appears to be an explosion to people who watch it 13.7 billion years later. That is the bang of the Big Bang.

Such an exponential growth is caused by[2] a constant energy density ρ, so when the linear size of the universe grows 10^{25} times or more, the energy in the universe grows like its volume by 10^{75} times or more. Whatever energy the universe possessed at the beginning of the classical Big Bang, there was at least 10^{75} times less at the beginning of inflation. It is in this sense that '*the universe started out from nearly nothing*.'

Exponential inflation was also proposed by Andrei Linde, who however was not aware of the classical Big Bang problems. The inflation theory proposed by Guth was subsequently improved by Linde, Andreas Albrecht and Paul Steinhardt, and many others.

For inflation we need a constant energy density ρ like a vacuum for some time but eventually it must decay. This kind of 'temporary vacuum' is known as a *false vacuum*, and it must live long enough to enable the universe to grow 10^{25} times. There are many suggestions and many models for the false vacuum but at present we do not know which of them is correct.

Inflation solves the flatness problem for the following reason. Both the kinetic and the potential terms of the Friedmann equation have grown a factor of 10^{50} over a short time, so that whatever $-k$ was before the inflation, it has to be completely negligible compared to each of these two terms at the end of inflation. In this way the universe became very flat at the end of the inflation, which is the beginning of the *classical* Big Bang era. It is this flatness requirement that determines the growth factor to be more than 10^{25}.

Inflation also solves the horizon problem, because the pre-inflated region that eventually develops into the whole observable universe is *very* small. As long as there is sufficient time for light to get across this pre-inflated small region before it grows 2.718 times,[3] then at the beginning of the classical Big Bang the observable universe was already homogeneous and uniform throughout, in which case an isotropic CMB is no longer a mystery.

Before inflation, the observable universe was very small, with very little energy in it. In this sense we say '*the universe started out from nearly nothing.*' In the bank analogy of Chap. 12, the initial balance of your local bank account was very small, but presently you discover you have 10^{75} times more. We will worry where it comes from later, but let us first get a feeling for how much money that represents.

Suppose your initial deposit is one cent in U.S. dollars. Just one cent. Now you have 10^{73} dollars. How much is that?

According to Forbes, the richest person in the world in 2005 was Bill Gates, whose net worth that year was reported to be 46.5 billion dollars. The estimated world population is roughly 6.5

billion, so if everybody on earth were as rich as Bill Gates, the total wealth of everybody put together would be about 3×10^{20} dollars. There may be something like 10^{21} stars in the whole universe, so if every one of them has an earth as rich as that, then the total amount of money in the universe is *only* 3×10^{41} dollars. This is *absolutely nothing* compared to the 10^{73} dollars in your account. Now I hope you realize how *enormously large* this minimal inflation factor is.

We can also make an estimate of how much seed *energy* is needed initially for it to blossom into our observable universe. This depends on exactly how many times the universe has inflated, and it also depends on the reheating temperature, i.e. the initial temperature at the beginning of the classical Big Bang universe.

If F is the inflation factor for the linear size, which is at least 10^{25}, and if T_{init} is the reheating temperature, then the seed energy needed before the inflation is[4] estimated to be $S = 4.5 \times 10^{54}\, F^{-3}\, (T_{\text{init}}/0.74 \text{ eV})$ kg. Since we do not know the reheating temperature or the precise inflation factor, we do not know what S must be. There is some indication that matter as we know it can emerge only when $T_{\text{init}} > 10^9$ GeV (Chap. 18), and we already know that $F > 10^{25}$. Taking the minimum values of both of these two factors, we get $S = 6 \times 10^{-3}$ kg $= 6$ grams. We will get different S by taking different values of F and T_{init}.

Either from this energy estimate, or the money analogy, I think it is perfectly justified to declare that *the universe started out from nearly nothing.*

Let us now ask where all this energy of the present-day universe comes from, and what is the origin of this ultimate free lunch?

To figure that out, let us return to the Friedmann equation of the last chapter. This equation governs the evolution of the universe, from pre-history to the present day. It is derived (note 15[3]) by putting a non-relativistic 'test object' in the universe, letting it move with the expansion of the universe, and requiring its total energy (rest plus kinetic plus potential energies) to be a constant. This test object can be identified with one of the raisins in the raisin bread model of the universe (Fig. 19). If we further assume the dough of the raisin bread to be the vacuum, playing essentially no role from inflation to the matter era, and the non-vacuum energies of the universe to be carried entirely by the raisins, then the only way the total energy of the universe can change with time is for the total number of raisins within the observable universe to change.

Let us first examine the matter-dominated era, where the energy density ρ decreases like the inverse volume, so that the total energy of the universe is a constant. This requires the number of raisins to remains constant as the bread rises in the oven, like real raisin bread does. Furthermore, we may consider the raisins to be the (non-relativistic) particles in the universe, so the conservation of the number of raisins agrees with the conservation of baryonic and leptonic numbers, because in the matter era anti-particles are no longer present and neutrons will no longer undergo beta decays (Chap. 19).

In the radiation-dominated era, the energy density ρ decreases like the inverse fourth power of the scale factor, so the total energy of the universe decreases like the inverse scale factor. As remarked before, the lost energy is utilized to overcome the pressure present in this era, so we know where it goes. In terms of the (relativistic) particles present in this era, their total number is conserved, but their individual energy decreases with the inverse scale factor, causing the total energy of the universe to decline in the same way. The decline of the energy of individual relativistic particles can be viewed either as a result of the stretched wavelength as the universe expands, or as the energy spent in pushing other relativistic particles aside when it moves out with the expanding universe.

In terms of raisins, this energy decrease of the universe causes the number of raisins to decline in the same way. Let us see how this comes about. Unlike the matter-dominated era, we can no longer identify the raisins with the particles in the universe, because the particles in the radiation-dominated era are relativistic, and the raisins are non-relativistic. However, we may consider a raisin to be an aggregate of several relativistic particles, whose total energy in their center-of-mass is mc^2. As time goes on, the decline of energy of individual particles makes it necessary to have more relativistic particles in the aggregate to attain the same rest energy mc^2; hence, the number of raisins must decrease if the number of relativistic particles is to stay constant. In this case, raisin bread is no longer a very good model of the universe in this era — unless we include a rat to eat the raisins!

Now we come to the inflationary era, where the constant energy density implies a constant number *density* of the raisins. The original raisin bread model is viable only when we keep on adding raisins to keep its number density constant when the dough rises in the oven. But then we may ask, where do the raisins or energies come from? The formal answer is that it comes from a negative pressure (see note 15[3]). Just like the presence of a positive pressure in the radiation era can cause the total energy to be lost, the presence of a negative pressure can cause energy to be gained. That is a perfectly legitimate answer but since we have no experience with negative pressures it is hard to form a mental picture of it.

We can devise a different raisin-bread model which might make it easier to understand the situation at inflation. In this alternate model the number of raisins is fixed, all there already at the very beginning, though hidden from view.

In the original model, each raisin carries with it the rest energy, the kinetic energy, and the potential energy of a small part of the universe. In the alternate model, the bread does not rise and the raisins are at rest, so they can carry only the rest energies. Something else must be introduced to carry the kinetic and potential energies of the universe.

As in the original model, we assume the dough plays no role in what follows. To model the expansion of the universe, we continuously inject a yellow vegetable dye in the middle of the bread to enable its edge to spread out exponentially in all directions, somewhat like what is shown in Fig. 53. This dye carries the kinetic and potential energies of the universe, and marks a sphere with a radius proportional to the scale factor $a(t)$ at any time t. The kinetic and potential energies obey the Friedmann equation so this part of the energies is always conserved.

The advantage of this model is that the raisin density (energy density) is automatically constant without requiring additional raisins to be added from time to time. The 10^{75}-fold growth of the rest energy during inflation comes from the 10^{75} growth of the dyed volume. Since the raisins are always there, so are their rest energies. The observable (dyed) universe started out with almost no energy because the dyed volume was originally very small, and the un-dyed raisins were ignored. But if they were also counted, then all the 10^{75} fold of energies had always been there and the universe really did not start out almost empty!

This is a bit like the bodhi tree and the mirror stand in Hui Neng's poem. They are there all the time, but his Buddha nature and his Buddha mind chose to ignore them for a reason. In that sense the emptiness of the primordial universe and the emptiness in Buddhism are somewhat similar, in that in some sense neither is really empty.

The energy density during inflation is constant, and so is the dark energy seen at the present time. However, there is a big difference between the two.

First of all, inflation occurred at pre-historic times but dark energy appeared only very recently. Secondly, dark energy density is very small, being only 73% of the critical density, roughly the rest energy of four protons per cubic meter, but the energy density required for inflation is very much bigger. Thirdly, while the dark energy could very well be a vacuum energy, remaining constant forever, that is not possible for the energy density of inflation; otherwise inflation would go on forever and would not have stopped. The energy density in inflation must remain almost constant sufficiently long to allow the size to grow 10^{25} times or more, but at some point it must decay to zero to stop the inflation. This decay liberates a large amount of energy to be used to create particles and anti-particles, and to heat up the early universe to a high temperature (this is called *reheating*). We know for sure that the temperature has to be at least 10 MeV or else we would not have as much helium in the universe as we see today (Chap. 19). It is probably much higher, maybe higher than the billion GeV range (Chap. 18), but we do not know exactly what the reheating temperature is.

In spite of these differences, there are similarities because the universe grows exponentially both during inflation and when the dark energy completely dominates. A little mathematics shows that exponential growth occurs whenever the expansion rate da/dt

is proportional to a. If we plot the logarithm of da/dt against the logarithm of a, then we will get a straight line with a unit slope for both the inflationary and the dark-energy eras. The only difference between the two is that inflation occurs for small a and dark energy occurs at large a, so the straight line for dark energy stays to the right of the straight line for inflation.[5]

As remarked above, we do not know what precipitated inflation, and what happened before its onset. Let us sidestep the origin of inflation and ask whether the inflationary theory is really correct or not. Unfortunately, there is not yet an unequivocal answer, because no smoking gun has been found. However, there is much circumstantial evidence to indicate that the theory is correct. To start with, it solves the three problems of the classical Big Bang theory. At the time when it was proposed in the early 1980s, the universe was thought to be open. Now we know that dark energy is present in just the right amount to make the universe critical, within experimental error. This near flatness is what the inflationary theory predicts if it inflates long enough. There is no other scientific theory that so naturally explains why the universe is nearly critical — though the 'Anthropic Principle' does. In essence, the latter states that if the universe were not approximately flat, then we could not exist so we would not be here to ask the question.[4]

Then there is the microwave background radiation (CMB), which we shall discuss in more detail in the next chapter. As mentioned in Chap. 14, on top of a uniform and isotropic background lays a small fluctuation. According to the inflationary theory, this is the result of quantum fluctuation of the energy density in the inflationary era, when the universe was small enough to make quantum effects important. The inflationary theory also explains how this small fluctuation at the beginning of the universe can persist for 400,000 years until decoupling time to allow it to be observed in the CMB, and later, the same fluctuation is concentrated and amplified to form the stars and galaxies. The distribution of such fluctuations is controlled by the parameters of the cosmological model, so measurements of the distribution via CBM and galaxy distributions can give us the magnitude of these parameters that can be checked and finds agreements with other independent observations. No other theory has been proposed to be able to do all of these either.

Before we close let me mention *eternal inflation*. From what we have said up to now, inflation stopped when the false vacuum decayed into the real vacuum. If the quantum effect is taken into account, then dynamical evolution is probabilistic and not deterministic. In that case, a small piece of the false vacuum somewhere in space may happen not to decay at the same time. In that case, the universe around this second piece of space continues to inflate. Even if this second piece started out much smaller than the first piece, its continued inflation will soon make it as big as the first piece that has decayed. Now when the second piece of space decides to decay, a small third piece within it may continue to inflate and soon it will catch up in size with the first two pieces. This story may repeat itself over and over again, leading to an eternal inflation because some small piece of space, not connected to our own universe and probably not observable, may still be inflating.

However, that would happen only if all the branched-off universes continue to grow exponentially, an assumption which is not directly supported by any evidence.[5]

17

Cosmic Microwave Background Radiation

The CMB fluctuation seen in Fig. 37 turns out to contain very important information about the universe. It tells us that the universe is spatially flat to within 2% of observational uncertainty. It tells us that dark energy occupies 73% of the critical density, ordinary matter 4.5%, and dark matter 22.5%. At a present temperature of 2.725 K obtained from the uniform background, this translates to a ratio η of the number of nucleons to the number of photons in the universe to be about 6×10^{-10}. It tells us that the present Hubble expansion rate is 72 km/s/Mpc. This is often written as $h \times 100$ km/s/Mpc, with $h = 0.72$. It also tells us that decoupling occurred at a redshift of 1,089, and the universe entered the matter era at a redshift of 3,233. It tells us other things as well, some of which will be mentioned later.

These are very significant results. Not only have they given us much information about the universe, these numbers also agree with all other measurements within observational errors. This gives us great confidence in the correctness of the CMB theory to be described in this chapter, which in turn lends strong support to the validity of the theory of inflation because inflation is crucial to the CMB theory. In what follows, I will sketch in a qualitative manner how the CMB theory works.

In a broad brushstroke, what happens is the following. The tiny quantum fluctuation that occurred during the inflationary era is frozen and preserved after inflation. Later on it is defrosted and used to generate an acoustic oscillation (sound) in the cosmic plasma, whose intensity at decoupling time is depicted in Fig. 43, and whose shape and pattern at the same time is seen in Fig. 44. Both are plotted against the wave number $k = 2\pi/\lambda$, where λ is the wavelength. These distributions are affected by the parameters of the universe mentioned in the first paragraph, so by making a best fit to the observed data, these parameters can be deduced.

I will discuss this sequence of events in great detail later, one item after another, but for the sake of orientation here is a roadmap of what we are going to do. To start with, I will explain what a quantum fluctuation is. In the present context, it is simply an unavoidable but computable disturbance occurring in the inflationary era. I will then explain what an acoustic oscillation is and how a sound wave can be generated by the quantum fluctuation originated in the inflationary era. The strange thing is, this disturbance does not produce a sound wave right away. Rather, it gets frozen for a long time before it is defrosted and used to generate the sound. This strange behavior, a consequence of the expansion of the universe and the presence of the inflationary era, will be taken up next. The sound so generated cannot be heard in the vacuum of space, but its pattern at the decoupling time can be *seen* through microwave radiation because photons participate in the acoustic oscillations. They carry with them the oscillation pattern shown in Fig. 37, from which we can extract Figs. 43 and 44. Figure 44 depicts the oscillatory sound pattern at decoupling but it will take us a while to learn how to read it to extract information. Figure 43 reveals the frequency (actually the wave number) distribution of the sound waves. This frequency distribution differs from the frequency distribution of the primordial quantum disturbance (except at low frequencies) because the presence of matter, photons, expansion, and gravity modify the distribution. Such modifications also affect the sound pattern shown in Fig. 44. The amount of modification depends on the size of the cosmological parameters discussed in the first paragraph of this chapter, which is why these parameters can be obtained by fitting the graphs in Figs. 43 and 44.

After the photons are released at decoupling, gravity takes over to amplify the sound pattern into matter clumps, and eventually into galaxies and stars. Thus, galaxies originate from the sound wave and their distribution also reflects the sound wave distribution. The data from recent galaxy surveys incorporated in Figs. 44 (all except the green points) and 46 support this interpretation.

This then is the general roadmap for what we are going to visit in the rest of the chapter. I will explain the details and the views at each stop below.

Quantum Fluctuation

In daily life, we are used to describing several physical quantities together. For example, how fast a car is when it crosses a particular landmark, or what kinetic energy it possesses at a given time. Quantum mechanics tells us that such simultaneous specification of two quantities often contains a certain amount of unavoidable error. To be sure, these unavoidable errors are so small that we never have to worry about them in daily life, but they would become significant in microscopic systems such as atoms and nuclei. In that case we have to use quantum mechanics to handle them.

These unavoidable errors are known as *quantum fluctuations,* and their sizes are governed by the *uncertainty principle.* In its simplest form, the principle demands the product of errors in position and in momentum (mass times velocity) to be larger than $\hbar/2$, and the product of errors in energy and in time also to be larger than $\hbar/2$, where $\hbar$ is a very small quantity known as Planck's constant. It is so small that these quantum errors are completely unimportant in daily life. See note 16[1] for its exact value.

We have encountered the uncertainty principle before. In Chap. 11, it is used to reduce the effective attraction at short distances. This comes about because the uncertainty in position has to be small at short distances, so the uncertainty in momentum has to be relatively large. A large momentum gives rise to a large kinetic energy, effectively reducing the attraction provided by the negative potential energy.

We also used it in Chap. 12 to borrow energy for a short time. That is just another way to say that the product of uncertainties in energy and in time is of the order of $\hbar$.

In the context of inflation, when the tiny universe is in a false vacuum in the inflationary phase (Chap. 16), there is an unavoidable fluctuation in the energy density ρ which is responsible for generating the CMB fluctuation seen in Fig. 37. The amount of fluctuation will be discussed later under the section 'Power Spectrum.'

We move on now to the second stop in our tour, acoustic oscillations.

Acoustic Oscillation

When air is perturbed by snapping your fingers or by the vibration of your vocal cord, an (positive or negative) excessive pressure is generated locally. Like a spring being compressed or stretched, a reaction to this excessive pressure tries to return the disturbed air to its normal density. Again like a spring, when it reaches the normal density, inertia causes it to overshoot, ending up with an excessive pressure the opposite way. This sets up an oscillation between higher and lower densities, just like a disturbed spring vibrates about its undisturbed position. The oscillation generates a sound wave in the air, like the one shown in Fig. 16, whose pitch depends on the nature of the initial disturbance. Generally, the disturbance contains a mixture of many wavelengths, allowing it to generate a great variety of sounds by altering the mixture.

This is how a cosmic sound is generated as well, but in the cosmic plasma rather than the air. A cosmic sound is more complicated than an ordinary one because gravity is involved, because the plasma has many components, and because the universe is expanding. The initial disturbance that leads to sound generation originates from the inflationary era, but right after inflation this disturbance is frozen for a long time before it is taken out and defrosted to produce the sound. This strange delay is caused by the accelerated expansion of the universe during inflation, coupled with the decelerated expansion later. To see how that works, let us move on to the next viewpoint and discuss how physics is affected by the expanding universe.

Sound in an Expanding Universe

Sound in an expanding universe behaves quite differently to sound in the still atmosphere.

For one thing, the wavelength is not fixed. It is stretched by the expansion of the universe, with the same scale factor a.[1] To avoid the stretching and other expansion complications, it is better to deal with sound waves in the *comoving frame,* in which *comoving distance* and *conformal time* are used. The comoving distance, as defined at the beginning of Chap. 9, is frozen to be the present physical distance. It is also equal to the physical distance at a time t divided by $a(t)$. In the coordinate system where comoving distance is used, the velocity of light is no longer c, which causes a great inconvenience. To correct for that, a *conformal time* η is invented so that the speed of light and all other velocities remain the same as those measured with the physical distance and physical time.[2] In short, the conformal time η is the appropriate time to use in the frozen comoving frame.

Instead of the fixed comoving wavelength λ, in what follows it is often more convenient to use the comoving *wave number* k, defined to be $k = 2\pi/\lambda$. It is just the number of waves per unit distance multiplied by 2π. It is also equal to 2π times the frequency of the wave divided by the speed of sound. Since k and frequency are proportional, we will often refer to the distribution of k by a more familiar terminology as the frequency distribution.

Clearly, physical effects of the expansion cannot be washed away just by choosing a different coordinate system. The appropriate time scale to measure expansion in the physical world is the Hubble parameter $H(t) = a^{-1}(da/dt)$, so in the comoving frame it is aH.[3] Consider some physical event with a characteristic frequency in the comoving system. If this frequency is much larger than aH, then clearly it will not be much affected by the expansion, so the expansion can be ignored and the sound wave propagating in such an expanding universe is very much like the sound wave propagating in still air. If this frequency is much smaller than aH, then one would not notice the change caused by the disturbance in this expanding universe, so for all practical purposes the disturbance is frozen.

In terms of the comoving wave number, if k is much larger than aH/c, then the expansion of the universe can be ignored; if k

is much smaller than aH/c, then we can consider the disturbance to be frozen in time. This then is the important new phenomenon of sound propagation in an expanding universe: sounds with small wave numbers are frozen and preserved. This phenomenon does not exist in still air or a static universe where $H = 0$.

The disturbance is said to be outside the *Hubble horizon* if $k < aH/c$; it is said to be inside the (Hubble) horizon if $k > aH/c$. For our qualitative description, we will pretend that a disturbance is completely frozen once it is outside the Hubble horizon, and it is unaffected by the expansion once it is inside. Such behaviors are exactly true only deeply inside each region, but otherwise only approximately so. However, the approximation simplifies description and is good enough for our qualitative understanding of the physics.

In Fig. 42, a schematic plot of aH/c is given for various eras, together with the comoving wave number k of a typical sound wave. Since $aH = da/dt$ is just the rate of change of the scale factor, its exponential growth during the inflationary period forms a steep cliff on the left of the mountain-shaped curves. To the right of the mountain when we enter into the radiation and the matter eras, gravity and pressure cause da/dt to decline, but the decline rate is now only power instead of exponential in t. In other words, we find the mountain to have a far gentler slope on the right. When a disturbance with a fixed wave number k generated by the quantum fluctuation gets inside the mountain, it is not only out of view, so to speak, but it is actually out of the Hubble horizon and frozen. The smaller k is, the closer the fluctuation to the base of the mountain is, and the longer the disturbance out of view and out of the horizon is.

We are now ready to discuss the meaning of the observed sound pattern shown in Figs. 43 and 44.

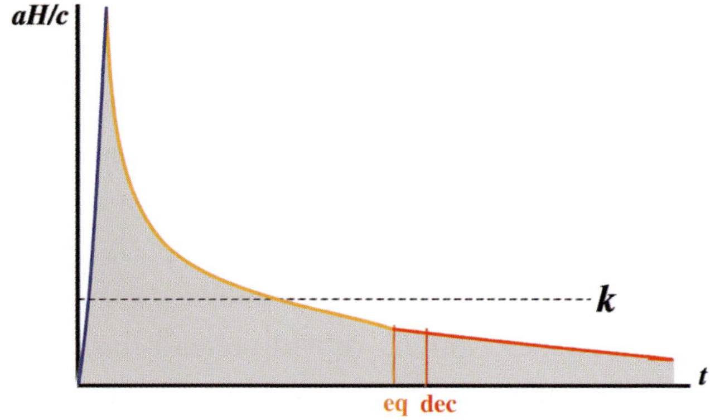

Figure 42: The colored curves are the aH/c values as a function of time t in the inflationary era (blue), the radiation era (orange), and the matter era (red). The two vertical bars indicate the time when the radiation era turns into the matter era (eq), and when decoupling occurs (dec). The wave with comoving wave number k is inside the Hubble horizon to the left of the blue curve and to the right of the orange-red curve. It is out of the Hubble horizon between the blue and the orange-red curves. Depending on the value of k, the sound wave can re-enter the horizon either in the radiation era or the matter era. It is important to note that this diagram just shows the qualitative behavior. Nothing is drawn to the proper scale.

Power Spectrum

Figure 43 shows the CMB *power spectrum,*[4] which tells us the intensity (loudness) of the cosmic sound as a function of the wave number k. Instead of k it is perhaps more intuitive to think of it as frequency, or the pitch. These are very low frequencies, way below our audible range, as the corresponding wavelengths are astronomically large, in the range of Mpc to thousands of Mpc.

As mentioned before, we see the sound through the CMB rather than hearing it. They are indicated by the green points in Fig. 43. These are points of low frequencies, or small k.

Those with larger k are not measurable with the present WMAP satellite but some of them will be accessible by the

Planck satellite to be launched soon. For now, we rely on galaxy distributions to obtain the data points for higher frequencies (points with other colors). As mentioned in the roadmap, after decoupling gravity amplifies the disturbances and pulls them together into matter clumps and eventually galaxies. Hence, the distribution of galaxy sizes reflects the distribution of sound wavelengths, and the latter can be inferred from the former.

The tone we hear (see) in Fig. 43 is not the original tone generated in the inflationary era. Ordinary matter, dark matter,

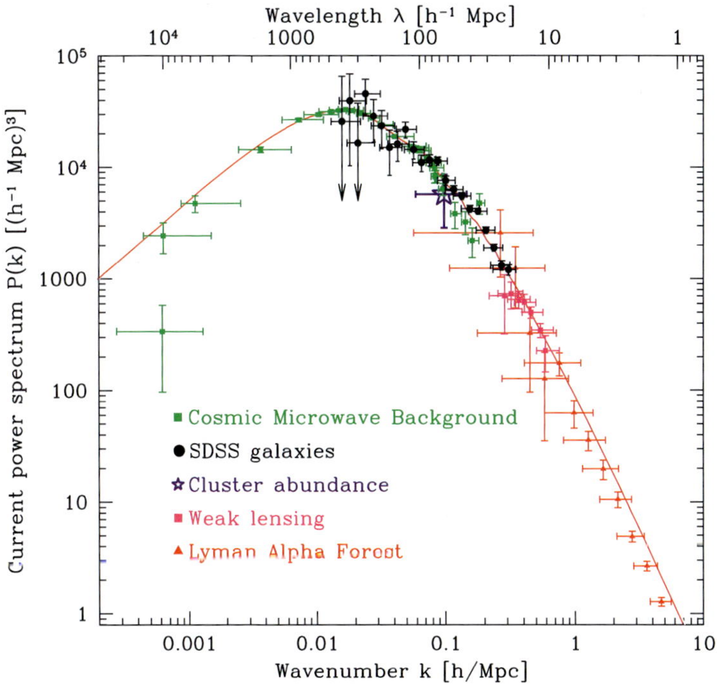

Figure 43: The power spectrum plotted as a function of k. For our purposes, we can take $h = 0.72$, and think of the power spectrum as the square of the observed amplitude. The green data are the CMB points, the others are obtained from galaxy surveys. The solid and dashed curves are theoretical curves computed from CMB and galaxy formation theories using two different sets of parameters. Courtesy of Max Tegmark/SDSS Collaboration.

expansion, as well as gravity act like acoustic materials in a room to modify the sound. Some frequencies are transparent to these materials, whereas others get partially absorbed. Since the universe has some 400,000 years to do it before we see the photons come to us at decoupling time, the frequency distribution of the sound that we see is no longer the same as the primordial distribution during the inflationary era.

However, recall from Fig. 42 that everything is frozen outside the Hubble horizon, when the wave number k is buried under the mountain. Thus, the acoustic materials can affect the sound pattern only to the right of the mountain. Moreover, I will explain later that the acoustic material is transparent to the low frequency sounds to the left of the peak in Fig. 43. Hence, that part of the distribution is still the primordial distribution in the inflationary era.

The original tone generated from quantum fluctuation in the inflationary era can be calculated. The wave number (frequency) distribution of the intensity turns out to be $(k^{4-3n}/M_P^2)(V/\varepsilon)^{1/2}$, evaluated at the inflationary-era horizon exit point $k = aH/c$, namely, at the point when the dashed line in Fig. 42 begins to enter the mountain from the left.

I will explain the symbols in the next paragraph. A comparison of the observed data in Fig. 43 to the left of the peak with this formula yields the '*COBE normalization*' $(V/\varepsilon)^{1/4} = 6.6 \times 10^{16}$ GeV, an important piece of information about the energy density in the inflationary era. It is truly amazing that such information about the very early universe 13.7 billion years ago can still be obtained.

In the formula above, V is related to the energy density ρ during inflation[5] by $V = \rho(\hbar c)^3$, so that $V^{1/4}$ has the unit of energy. In short, V is just ρ expressed in the unit $(\mathrm{GeV})^4$. H is the Hubble parameter in the inflationary era, which is equal to the exponent $\alpha = (8\pi\rho G/3c^2)^{1/2}$ of inflation discussed in Chap. 15.

Both V and H are approximately constant throughout the inflationary period. $M_P = (\hbar c/G)^{1/2} = 2.18 \times 10^{-8}$ kg $= 1.22 \times 10^{19}$ GeV/c^2 is the Planck mass (see note 16[1]). The parameter ε is one of two *slow-roll parameters* characterizing how slowly the false vacuum decays. The other one is called η (not to be confused with either the conformal time η or the nucleon-to-photon number ratio η). These slow-roll parameters must be small in order to allow the universe to stay at the constant density long enough to inflate to a minimum multiple factor of 10^{25}. The number n is called the *tilt*. It determines the k distribution, and is related to the slow roll parameter by $n = 1 - 6\varepsilon + 2\eta$. Its best-fitted value from the 3-year WMAP data is 0.958 ± 0.016, close to the '*scale invariant*' value of $n = 1$, and is consistent with the smallness of the slow-roll parameters. Note that Fig. 43 is a log–log plot. The straight line below the peak at about 0.01 h/Mpc represents a power dependence on k, agreeing with the spectrum predicted by inflation and allowing the tilt to be extracted from it.

If ε can be measured independently, then from the COBE normalization $(V/\varepsilon)^{1/4} = 6.6 \times 10^{16}$ GeV discussed above, $V^{1/4}$ can be computed to give us the density of the universe in the inflationary era. Unless ε is terribly small, $V^{1/4}$ is expected to be in the 10^{15} to 10^{16} GeV range.[6]

In principle there is a way to measure ε (see the discussion on polarization later), but in practice this measurement is very difficult. If ε is somehow known, then tilt n determines the other slow-roll parameter η.

I mentioned in the last chapter that the dark energy density is much smaller than the energy density in the inflationary era. Let us see how much smaller. The dark energy density, being 73% of the critical density, is 4×10^{-6} GeV/(cm)^3. In contrast, the energy density in the inflationary era is $\rho = V/(\hbar c)^3$. Taking $V^{1/4} = 10^{16}$ GeV, we get $\rho = 1.3 \times 10^{105}$ GeV/(cm)^3!

To summarize, what we have learned by comparing the small k part of Fig. 43 with the outcome of the quantum fluctuation calculation in the inflationary era are: (i) the wave number (frequency) distribution of the sound intensity obeys a power law in k, with a tilt parameter equal to $n = 0.958 \pm 0.016$; and (ii) the energy density V during inflation is constrained by the COBE normalization $(V/\varepsilon)^{1/4} = 6.6 \times 10^{16}$ GeV, where the slow-roll parameter ε is so far unknown but must be small. Otherwise inflation will not last long enough to produce an amplification factor of at least 10^{25} times for the size of the universe.

The Frozen Sound Pattern

Let us turn to Fig. 44,[7] which shows the frozen sound pattern seen at decoupling time.

This graph plots the temperature–temperature (TT) correlation, normalized in a certain way, against the multiple moment l. There is no way I can explain either of these two quantities precisely without a lot of mathematics, but I shall do it in an approximate way, preserving the essential physics.

The TT correlation is obtained by taking two points in the sky separated by an angle θ, multiplying together the amount of temperature fluctuation (namely, the difference of the actual temperature from the average temperature divided by the average temperature) at each of these two points, then averaging the product over the whole sky. For our purposes, we shall take that to be the square of the *displacement* of the sound wave observed at decoupling time, with a wave number related to θ.

The x-axis is the 'multiple moments' l. For our purposes, it is defined by $l = \pi/\theta$ with θ being shown as the angular scale on top of Fig. 44. This interpretation of l is approximately correct for large l, but not at a small l way below the first peak in the graph.

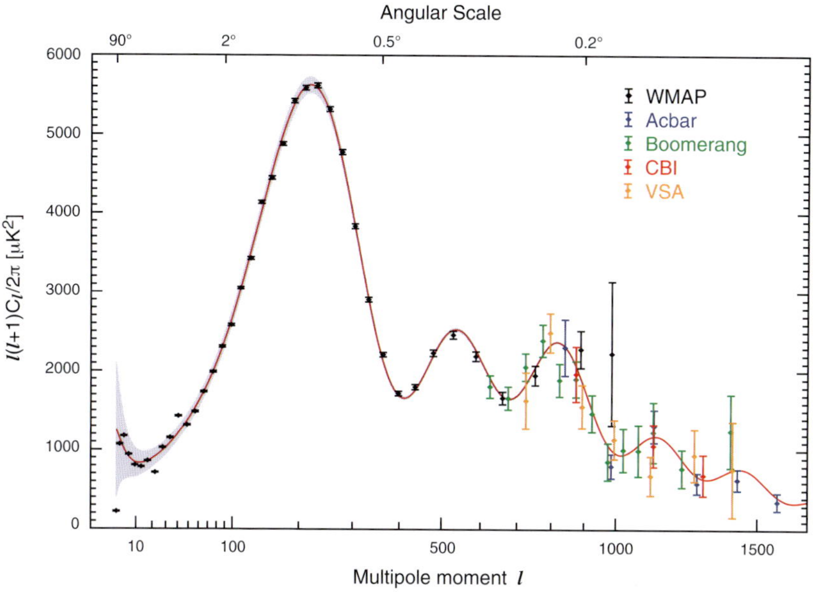

Figure 44: For our purposes, we can think of this as a plot of the square of the observed amplitude against the wave number k at the time of decoupling. The data with very small error bars on the left are from WMAP, the data at larger l (or k) are from other non-satellite experiments. The solid curve is computed from the CMB theory using the quoted cosmological parameters. Courtesy of NASA and the WMAP Science Team.

We may also think of l as a quantity proportional to the wave number k. In a flat space, the relation is $l = kc\eta_0$,[8] where η_0 is the age of the universe in conformal time. Since $l = \pi/\theta$ this formula gives a relation between the wave number k and the angle θ.

Both Figs. 43 and 44 plot against the wave number, but they differ in the following way. Figure 43 concerns the intensity of sound at a particular wave number, or frequency, and intensity is the square of the *amplitude*. Figure 44 concerns the square of the *displacement* at the time of decoupling (see the 'Wave Propagation' section in Chap. 6 for the difference between displacement and amplitude), and as such it gives an imprint of the frozen sound pattern at that particular moment.

The first peak in Fig. 44 occurs at $l = 220$. It corresponds to an angle of $\pi/220$ radians, or $180/220 = 0.8$ degrees. In terms of Fig. 37, this simply means that most of the spots are about one degree in size across the sky.

This peak also corresponds to a wave number $k = l/c\,\eta_0 = 0.02$ h/Mpc,[9] where $h = 0.72$ is the Hubble constant H_0 measured in units of 100 km/s/Mpc.

There are other peaks in Fig. 44 occurring at other angles θ, corresponding to wave numbers k at which the square of the acoustic wave displacement reach maximum values at the decoupling time. To see the origin of these peaks, let us first adopt a simplified approximation by forgetting ordinary and dark matter as well as gravity and expansion. This is the photon fluid approximation that will be discussed next. Acoustic peaks are produced at approximately the right locations within this approximation, but their heights are wrong because we have taken away the acoustic material. The right amount of matter and expansion must be included to render the right heights. This is how we can determine such parameters as the amount of matter and the expansion rate of the universe by figuring out the amount of acoustic material necessary to reproduce the observed sound pattern.

The main thing we learn in this section about the frozen sound pattern of Fig. 44 is that the first peak corresponds to an angular scale of about 0.8 degrees, which is the main size of the spots in Fig. 37. We also learn that there is an approximate relation $l = \pi/\theta = kc\eta_0$ between the wave number k, the multiple moment l, and the correlation angle θ.

Photon Fluid Approximation

In this approximation, we are dealing with a fluid of pure photons.

This fluid is hard to compress because of the presence of the large pressure exerted by the photons. The stiffness of the fluid translates into a large sound velocity $c_s = c/\sqrt{3}$, more than half the speed of light.

The quantum disturbance originated from the inflationary era generates oscillatory sound waves of varying wave numbers. Different wave numbers give different periods of vibration and hence reach different displacement levels at the conformal time η_* of decoupling. If we plot the displacement at this time η_* against the wave number k, we get the red curve in Fig. 45. If we plot the displacement square, we get the blue curve.

The successive peaks in the blue curve are separated by a wave number $\Delta k = \sqrt{3}\pi/c\eta_* = \pi/c_s\eta_*$.[10] In terms of the l variable, these correspond to peaks separated by a multiple moment $\Delta l = \sqrt{3}\pi\eta_0/\eta_*$. Putting in $\eta_0/\eta_* = 50$,[9] this gives a value Δl = 274. In particular, the first non-zero peak should occur at l = 274, surprisingly close to the observed peak at l = 220 in Fig. 44, considering the crudeness of the photon-fluid approximation.

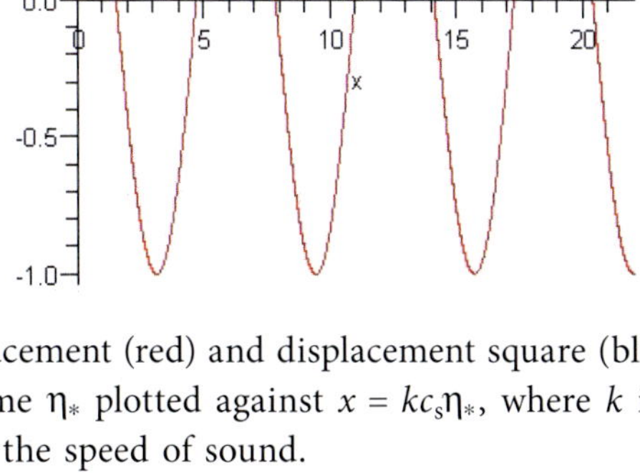

Figure 45: Displacement (red) and displacement square (blue) of a sound wave at decoupling time η_* plotted against $x = kc_s\eta_*$, where k is the wave number and $c_s = c/3^{1/2}$ is the speed of sound.

There is a more direct and more intuitive way to understand the position of the main (first) peak in Fig. 44. The comoving distance that a pulse of sound has traveled from the beginning to the decoupling time is $c_s\eta_*$ a distance known as the *sound horizon*. This is the length of correlation at decoupling time and therefore where the main peak in Fig. 44 lies. This distance subtends an angle $\theta = c_s\eta_* / c\eta = \eta_* / \sqrt{3}\eta$ (see Fig. 54) in the sky, and hence a multiple $l = \pi/\theta = \sqrt{3}\pi\eta_0 / \eta_*$, as obtained in the last paragraph.

This conclusion assumes flat-space geometry. It is no longer valid if the space is curved, as in Fig. 21. For a universe with a positive curvature, the first peak is located at a smaller l than the one calculated assuming a flat space, and for a universe with a negative curvature, it is located at a larger l.[11] Thus, from the location of the first peak one can determine whether the universe is closed or open, and therefore the amount of dark energy to make that happen. These features remain to be true even if we take into account matter and expansion effects.

In summary, the frequency spectrum of the sound intensity in the photon fluid approximation is given by the blue curve of Fig. 45. Compared to the observed result depicted in Fig. 44, the location of the peaks comes out approximately correct (except the one located at $k = 0$ in Fig. 45) in this approximation, but not their heights. To improve on the locations and to get the right heights, we have to restore the presence of matter, gravity, and expansion of the universe.

The Effect of Matter

There are many differences between the peaks in Fig. 44 and the blue peaks in Fig. 45. There is also a marked difference between the observed power spectrum in Fig. 43 and the primordial power spectrum proportional to a power of k. These differences are due to the presence of matter and gravity, as well as expansion and other smaller effects. It is from such differences that the amount of ordinary and dark matter as well as other cosmological parameters is determined. Once the matter content is known, the amount of dark energy present can be determined from the flatness of the universe.

Let us now discuss some of these corrections to the photon fluid approximation.

First of all, the minima in Fig. 45 between the peaks always reach down to zero, but not in Fig. 44. This is because realistically the initial oscillation does not start from rest. The resulting Doppler shift makes an out-of-phase contribution which fills in the zeros, and it also shifts the location of the peaks by a small amount.

Secondly, the peaks in Fig. 45 are of equal height but those in Fig. 44 decrease for higher l. Moreover, the peak at $k = 0$ in Fig. 45 is absent in Fig. 44. These differences are due mainly to the gravitational and matter effects explained below.

Without gravity, the compression phase and the stretching phase of the oscillation are symmetrical. In the presence of an attractive gravity, there would be more compression and less stretching, and hence the amplitude square of the compression (odd) peaks are higher and the stretching (even) peaks are lower than those shown in Fig. 45 (the peak at $k = 0$ in Fig. 45 is counted as the zeroth peak). This partly explains why the second peak in Fig. 44 is lower, but it does not explain why the third peak is lower still. That effect comes from the direct presence of ordinary and dark matter.

Note that there is a difference between ordinary matter and dark matter because the former interacts with photons and is a part of the oscillating plasma, but the latter is not coupled to the photon so it contributes only to the building up of gravitational potential.

The presence of ordinary matter provides an added inertia to the plasma oscillation. It also contributes to the energy density but not the pressure, so it slows down the velocity of sound c_s and shortens the sound horizon $c_s\eta_*$. These effects become more and more important as we get closer and closer to the matter-dominated era.

In contrast, dark matter does not contribute directly to oscillation, but it contributes to gravity which compresses matter. In the matter-dominated era, this tendency to increase density is almost exactly balanced by the expansion of the universe which tends to decrease it. As a result, density, sound amplitude, and gravitational potential remain fairly constant throughout the matter era. Since disturbances with a low k re-enter the horizon directly into the matter era (see Fig. 42), the observed sound amplitudes remain constant so they are the same as their primordial initial amplitudes. In other words, the intervening universe is actually transparent to low frequency sound waves. This explains why at small k the observed power spectrum in Fig. 43 is the same as the primordial power spectrum, proportional to a power of k.

It also explains the absence of a peak at zero l in Fig. 44, although it is present in the simplified model of Fig. 45 at $k = 0$. When the disturbances re-enter directly into the matter era, pressure is absent so there is no oscillation, and the observed amplitude remains approximately constant. This is why an oscillation peak is absent for small k when gravitational and matter effects are included.

For disturbances with higher k which re-enter into the radiation-dominated era, the presence of photon pressure prevents a gravitational collapse, but the expansion of the universe is still present to dilute the local density and gravity, and to reduce the observed amplitude, thus causing the peaks in Fig. 44 to decrease with increasing *k*, or *l*.

This also causes the power spectrum in Fig. 43 to drop away from the primordial power spectrum proportional to a power of *k*.

Indeed, the linear dependence in Fig. 43 stops close to the peak, at *k* about 0.01 *h*/Mpc. Here, $h = 0.72$ is the Hubble constant H_0 measured in units of 100 km/s/Mpc. This agrees with the explanation above because it is equal to the aH/c value at the boundary between the radiation and the matter eras.[12] For values of *k* larger than that, the disturbance re-enters in the radiation era, thus causing the observed power spectrum to drop.

Finally, there is another effect which causes the magnitude of the observed amplitude to decrease at large *k* or *l*. This effect is important for large *l* so it should definitely be taken into account for the Planck satellite data. I am referring to the leakage of photons. Up to this point we have assumed the photons to be completely trapped within the plasma and oscillate with it in unison. However, when *k* is large and the oscillation fast, photons cannot be bounced fast enough between the charged particles in the plasma to keep in sync, and a leakage will occur. This weakens the pressure and hence the oscillation amplitude.

Matter Oscillation

The evidence for acoustic oscillation shown in Figs. 37, 43, and 44 is gathered from the CMB data of light. By now evidence for acoustic oscillation has also been seen in galaxy density correlations. If you imagine a pulse of sound being sent out at the beginning of the universe from any point, then by decoupling time η_*, this plasma pulse has traveled a comoving distance equal to $c_s\eta_*$. Since this pulse contains ordinary matter, it shows

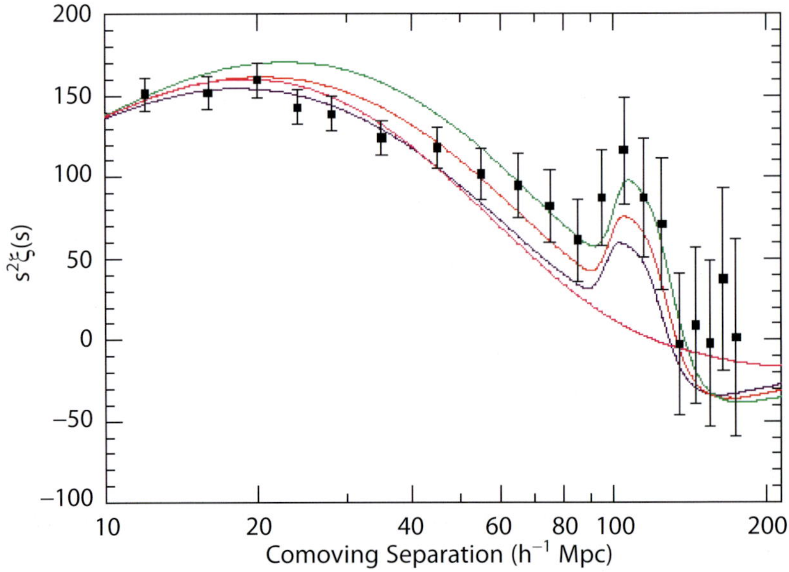

Figure 46: Correlation of galaxies shows a peak for galaxies about 100 Mpc/*h* apart. The magenta curve shows what is expected if there is no sound oscillation, and the other color curves show what is expected with sound oscillation and different amounts of matter in the universe.

up as a peak at about 120 Mpc/*h*[9] in the galaxy density–density correlation graph of Fig. 46.[13]

Polarizations

Light comes in two 'polarizations,' right-handed and left-handed (Chap. 11), but it is more convenient here to combine them and reclassify into either the 'E-type' or the 'B-type.' Since we have not discussed light polarizations at length, I will not be able to describe what they are, except to mention that the E-type polarization has been detected by WMAP and the result is consistent with what we know so far, but B-type polarization is much smaller and has not yet been detected. So far we have concentrated on sound waves generated by quantum density fluctuations in the early universe, but gravitational waves during inflation can also cause CMB fluctuations. Moreover, it contributes to the B-type polarization whereas density fluctuations do not. Thus, in principle, one can deduce the amount of gravitational waves in the early universe by measuring the B-type polarization. The ratio of the gravitational wave amplitude to the density amplitude is proportional to the slow-roll parameter ε so ε is known if the B-type polarization is measured.

Evidence for Inflation

The seed of CMB fluctuation lies in the inflationary era. Although small, it nevertheless requires the presence of a constant high energy density, sufficiently long to achieve the COBE normalization $(V/\varepsilon)^{1/4} = 6.6 \times 10^{16}$ GeV. If inflation were absent, there would be no cliff to the left of the mountain in Fig. 42. In that case, disturbances that generate the CMB fluctuation must originate from the right of the mountain. Since different wave numbers re-enter the horizon at different times, it is hard to imagine how their fluctuations can be related like those shown in Figs. 43 and 44. Moreover, since the presently measured wave numbers typically emerge from the mountain around radiation-matter equality, at which time the energy scale is very low, it is also hard to get a large enough fluctuation to satisfy the COBE normalization. For these reasons, the successful prediction of the CMB theory on the structures of Figs. 43, 44, and 46, as well as the successful reproduction of cosmological parameters obtained by other means, is a strong indication for the correctness of the inflationary theory.

18

Emergence of Matter

The Difficulty of Having Enough Matter in Our Universe

At the end of inflation, the false vacuum decays to produce particles at a high temperature. Since the decay comes from a false vacuum, only particle/anti-particle pairs are produced, so the net leptonic number (L) and the net baryonic number (B) (see Chap. 11) of the universe must both be zero.

Our universe seems to contain no anti-particles. Although the number of nucleons is far fewer than the number of photons, with a ratio $\eta \cong 6 \times 10^{-10}$, it is already much larger than we have a right to expect in a universe with no baryonic number. If the absence of anti-nucleons comes about because they have all been annihilated, then we should be left with no nucleons either. Even if the missing anti-nucleons are hidden in a corner of the universe not yet discovered, the ratio still cannot be larger than $\eta \cong 10^{-18}$. This comes about in the following way.

In the beginning when plenty of thermal energy was available, whatever was annihilated would be produced again, maintaining a thermal equilibrium. As the universe expanded and cooled, there came a point when thermal energy was so diminished that this pair production was no longer possible. Once a pair was annihilated the particles were gone forever, causing the number of matter and anti-matter particles to decline steadily. Nevertheless, this decrease would not continue forever because of the expansion of the universe. Expansion reduced the density of particles, thus causing the annihilation rate to drop. When the annihilation rate dropped below the expansion rate of the universe, annihilation stopped because it became difficult for an anti-particle to find a partner to annihilate with. This *freeze-out* of nucleons happened around a temperature of 20 MeV. The leftover number of nucleons (and anti-nucleons) at that temperature can be calculated, and this gives a η of about 10^{-18} quoted above, more than a hundred million times smaller than what is really observed today.

To obtain enough nucleons, we must search deep into our knowledge of particle physics for a possible mechanism to generate more matter than anti-matter particles. Such a mechanism is known as *baryogenesis*. In spite of the name, it generates not only extra baryons but also extra leptons. In this chapter, I shall describe one of the most popular baryogenesis mechanisms, known as *leptogenesis*. It is a two-step process in which extra anti-leptons over leptons are first generated, then some of these anti-leptons are converted into baryons through a '*sphaleron*' mechanism, described later.

It turns out to be very difficult to find particle physics processes that can accomplish the job. Andrei Sakharov pointed out in 1967 that in order for a net amount of matter particles to be created in the universe, three conditions must be satisfied. These Sakharov conditions are:

(i) Baryonic or leptonic number conservation must be violated.
(ii) C and CP invariance must be violated.
(iii) Thermal equilibrium must be violated.

From our present knowledge of particle physics, none of these conditions are very easy to meet, which is why the generation of matter is such a difficult task. In fact, with the present knowledge of particle physics that has been experimentally confirmed, there is no mechanism that can obey these three conditions and produce the desired η. Every viable mechanism involves some new physics, and some new particles. Leptogenesis is considered by many to be the most favored mechanism because there is good circumstantial evidence hinting at the presence of these new items needed in that process.

I have not yet discussed what the Sakharov conditions mean, why they are necessary, and why they are so difficult to satisfy. To do so meaningfully, we must know where present particle physics stands, what it allows to happen and what it forbids. Particle physics is a vast field and cannot be thoroughly reviewed in a book like this, so I will only pick those topics that are essential to our understanding of the Sakharov conditions in general and the leptogenesis mechanism in particular.

The present particle physics theory is known as the *Standard Model* (SM for short), and this is what we are going to take up next. After that we will be ready to discuss what the three Sakharov conditions mean, why they are necessary and why they are so difficult to satisfy. Finally, we will come to the leptogenesis process itself to see how the right amount of matter can be generated.

For readers who do not want to follow these details, they can skip right to the last section on leptogenesis. A general description of the basic ideas of leptogenesis is sketched at the beginning before going into further details that uninterested readers can skip.

The Standard Model

Fermions

We discussed in Chap. 11 the common constituents of matter: protons, neutrons, electrons, neutrinos, and their anti-particles. These are all fermions. We also discussed the photon, which is a boson.

By the 1960s, we learned that each nucleon was made up of three *quarks.* Quarks are funny objects: they cannot exist in isolation, but they can exist in aggregates of three in the form of nucleons. A quark and an anti-quark can also bind to form a boson.

Since there are two kinds of nucleons, the protons and the neutrons, we need two kinds of quarks. They are called the *up quark* (u) and the *down quark* (d), respectively. A proton has two u quarks and one d quark (uud), and a neutron has two d quarks and one u quark (ddu). Since a proton carries a positive unit of electric charge and a neutron is electrically neutral, it is not hard to figure out that an up quark carries an electric charge +2/3 and a down quark carries an electric charge −1/3. Moreover, both kinds of quarks must carry a baryonic number $B = 1/3$ because three of them makes a nucleon.

For reasons I do not want to explain because it really has little bearing on what we are interested in here, each quark actually comes in three different '*colors.*' The name 'color' is just a fancy name for a new kind of charge that has nothing to do with what we usually associate with this word. What we need to know is that the three quarks making up a nucleon are in three different colors.

With quarks replacing nucleons, the fundamental fermions are now the up and down quarks (u,d), the electrons and the neutrinos (e^-, ν_e), and their anti-particles. The minus sign in the upper right corner of the electron symbol e indicates its electric charge, and the subscript e for the neutrino symbol ν reminds us that this is the neutrino associated with the electron.

In 1936, Carl Anderson discovered a new particle in cosmic rays. In every respect that new particle looks like an electron, except that it is more than 200 times heavier, and it is unstable, with a lifetime of about 2×10^{-6} seconds. This particle is now called the muon, denoted by the symbol μ^-. The discovery of this particle was so unexpected, and its existence seemed to be so 'unnecessary' that it prompted the famous physicist I. I. Rabi to exclaim: "Who ordered that?" To this date, we still do not know the answer to this question.

In subsequent years, many other fundamental but unstable fermions have been discovered. We now know that the four fundamental fermions that we are familiar with, the up and down quarks, the electron and the neutrino, together with their anti-particles, all have two heavier unstable replicas, making a total of three *generations* (or three *families*) of fundamental fermions. Their names are shown in Fig. 47 below.

Gauge Bosons

The fermions interact with one another through four kinds of fundamental forces. Gravity is always there, but between these tiny fundamental particles it is so feeble as to be completely negligible. What remain are the strong, the electromagnetic, and the weak interactions. In the SM, there is an intimate connection between the last two so they are often mentioned together as the electroweak interaction.

These interactions can lead to particle decays and reactions. Essentially, any conceivable process not forbidden by conservation laws may occur, though some more rarely than others. Like a chemical reaction or an arithmetic equation, a reaction can be manipulated following some rules to become another possible reaction. For example, if A, B, C, D are four particles and the reaction $A+B \to C+D$ occurs, then generally the reactions $C+D \to A+B$, $B \to C+D+\bar{A}$ can both occur, provided they are not forbidden by energy conservation or other conservation laws. Here and below anti-particles are denoted by a bar on top. In general, the rules for allowed manipulations are: the reaction direction can be reversed, and a particle can be moved to the other side of the arrow provided it is changed into its anti-particle.

Everyone of these interactions, or forces, is mediated by the emission, absorption, formation, or decay of spin-1 bosons, known as the *gauge bosons*. The gauge boson for electromagnetic interaction is the *photon* (γ); that for the strong interaction is the gluon (g), which comes in 8 different colors; and those for the weak interactions are the $W^{\pm}$ and the Z^0 bosons. Each emission, absorption, formation, or decay process is known as a *vertex*. All interactions and decays can be decomposed into elementary vertex processes. For example, the electromagnetic scattering of two up quarks, $u + u \to u + u$, can be decomposed as $u + u \to (u + \gamma) + u \to u + (\gamma + u) \to u + u$. Their strong scattering can be decomposed in the same way with gluons g replacing the photons γ, and their weak scattering with the Z^0 boson replacing γ. Finally, the beta decay of a down quark, $d \to u+e^- +\bar{\nu}_e$, which causes the beta decay of a neutron (ddu), is mediated through W^- emission and decay: $d \to (u+W^-) \to u+(e^- +\bar{\nu}_e)$.

The meaning of C and CP in the second Sakharov condition will be discussed next. These are operations on a physical quantity or a physical process.

C, P, T Operations

C stands for *charge conjugation.* It is an operation that turns all particles into their anti-particles.

P stands for *parity.* It reverses the spatial coordinates, changing (x, y, z) to $(-x, -y, -z)$, but leaving time t untouched. Since velocity is the time rate of change of spatial position, and momentum is mass times velocity, both velocity and momentum also reverse their signs under a parity operation. Spin (see the end of Chap. 11) is an angular momentum, involving a product of position and momentum; hence, it does not change under P. Helicity, which is the projection of spin along the momentum direction, changes a sign.

T stands for *time reversal.* It reverses the time t, changing t to $-t$, without affecting the spatial positions (x, y, z). Hence, velocity and momentum both change a sign, and position does not. This causes orbital angular momentum, and hence spin, to change a sign, but helicity does not.

Since T reverses time, it also reverses a reaction. The reaction $A + B \to D + E$ becomes $D + E \to A + B$, with all the momenta and spins of the backward reaction reversed.

For a long time it was believed that all the fundamental interactions are invariant (unchanged) under any of the C, P, and T operations. That means every physical law and every physical quantity should be the same under any of these operations. In that case, Sakharov's second condition can never be satisfied.

The invariance is still believed to be true for strong and electromagnetic interactions, but not for the weak interaction. In 1957, Tsung Dao Lee and Chen Ning Yang discovered that parity P was violated in the weak interaction. Subsequently, it was found that C was also violated in the weak interaction, but the combined operation CP seemed to be conserved. This thought persisted until

1964 when James Cronin and Val Fitch discovered a rare process in the weak interaction which violated CP. With that discovery, we know the C and CP violating processes demanded by Sakharov's second condition do already exist in the SM. This makes it more likely that the second Sakharov condition can be fulfilled if new physics is introduced.

Every candidate theory of particle physics is believed to obey the *CPT theorem*, which requires the theory to be invariant under a combined C, P, and T operation. This operation changes a reaction $A+B \rightarrow D+E$ to the reaction $\bar{D}+\bar{E} \rightarrow \bar{A}+\bar{B}$, with all spins reversed.

Higgs Bosons

Other than the gauge bosons and the fundamental fermions, there are also four spin-0 bosons in the SM called the *Higgs fields*. They are (ϕ^+, ϕ^0), and their anti-particles $(\phi^-, \bar{\phi}^0)$. Their main mission in life is to cause masses to be generated.

Ordinary matter usually find itself in one of several phases: gaseous at high temperatures, liquid at moderate temperatures, and solid at low temperatures. The particles in an electroweak theory also find themselves in two possible phases: the *symmetric phase* at high temperature, and the *Higgs phase* at low temperature. The transition (the *electroweak phase transition*) occurs at a temperature of about 100 GeV.

In the symmetric phase, all particles are massless, and all the spin-1 gauge bosons have only two polarizations each, just like the photons.

In the Higgs phase, the self-interaction of Higgs fields binds them together to form something like a liquid which fills all space. In other words, the vacuum (lowest energy state) in this phase is filled with a liquid of Higgs fields. Since the vacuum has no charge, this liquid must consist of only the neutral Higgs fields. This liquid is known as the *Higgs condensate* or the *Higgs sea*. The universe at the present temperature is in the Higgs phase, so it might seem strange that we are not consciously aware of the Higgs sea although we are immersed in it and we walk in it. That is because we are born in it so we get used to it, just like a fish in water is (probably) aware of the water only when you take the water away from it. Similarly, we would not survive if the Higgs sea were removed.

According to the SM, particles have non-zero masses precisely because they are moving in the Higgs sea. Just as a boat in water is slower than a hovercraft because it receives more drag from water, particles moving in the Higgs sea also experience a drag which slows them down. Originally, in the symmetric phase, these particles are all massless, so they travel at the speed of light. Now, dragged and slowed down, they must travel at a slower speed and hence they must have a mass, for according to the theory of special relativity massless particles must travel with the speed of light.[1] The only particles that can escape this fate of gaining weight are those that do not interact with the neutral Higgs fields. Any particle with a weak interaction has to interact with the Higgs fields, so that leaves only the photons and the gluons massless in the Higgs phase of the present day.

The three weak gauge bosons $W^{\pm}$, Z have weak interactions so they all become massive. Being spin-1 particles, to be massive they must carry three spin orientations, but in the symmetric phase each of them carries only two. Thus, they must each incorporate an extra polarization somewhere, and they do so by taking over three Higgs fields as their own. The $\phi^{\pm}$ are incorporated into $W^{\pm}$ to make them massive and the orthogonal mixture of ϕ^0 and $\overline{\phi}^0$ is incorporated into Z to make it massive. The excitation of the original mixture of ϕ^0 and $\overline{\phi}^0$ in the Higgs sea, analogous to the waves in our ocean, shows up as a new neutral particle called the *Higgs boson*. Of the four Higgs fields in the symmetric phase only this one survives as an independent particle in the Higgs phase. This particle will be eagerly sought after in the LHC (see Fig. 32).

Elementary Particles

Figure 47 gives a summary of all the elementary particles in the SM, as well as the heavy right-handed neutrinos *N* whose existence is only inferred from the tiny mass of the left-handed neutrinos ν (see the 'Neutrinos' section below).

u	c	t	g
d	s	b	γ
e^-	μ^-	τ^-	$W^\pm$
ν_e	ν_μ	ν_τ	Z
N_1	N_2	N_3	$\phi^\pm, \phi^0, \overline{\phi}^0$

Figure 47: Elementary particles in the SM plus the heavy neutrinos.

The first three columns are the matter particles (spin-½ fermions); the first generation in the first column, the second generation in the second column and the third generation in the third column. Quarks are shown in red, to remind us that each of them comes in three *colors,* and leptons are shown without color. The light neutrinos ν and the heavy neutrinos *N* are both their same anti-particles, which is why they are shown in grey whereas the charged leptons, different from their anti-particles, are shown in solid black. Roughly speaking, two grey particles make up a solid black particle. The fourth column contains the bosons in the SM, with the spin-1 gauge bosons shown in light green, and the spin-0 Higgs bosons shown in orange.

I remarked above that only particles without weak interactions can hope to remain massless in the Higgs phase. Actually, there is another way to be massless, and for a long time that was thought to be the case for neutrinos. Neutrinos do have weak interactions, but they do not interact with the Higgs bosons if a neutrino is distinct from an anti-neutrino. To understand why that is the case let us study how spin-1/2 fermions gain their masses.

Fermion Masses

Let ϕ be any of the four Higgs fields, and a, b be two fermions. When we say a interacts with the Higgs, all we mean is that the reaction $a \rightarrow b + \phi$ can take place if energy is available and conservation laws are not violated. According to the uncertainty principle, energy can always be borrowed for a short time period, so the availability of energy is not an issue. Whether b is the same as a or not is governed by charge conservation. Moreover, if a is a lepton (quark), then b must also be a lepton (quark). If the Higgs field is neutral, a and b may be the same. If it is charged, they must be different; an example of the latter kind is $e^- \rightarrow \phi^- + \nu_e$. The probability of the reaction $a \rightarrow b + \phi$ occurring will be called the *coupling*, an abbreviation for what is usually known as 'the absolute magnitude square of the coupling constant.' In the SM, the coupling to the neutral Higgs fields is proportional to the square of the fermion mass.

If the reaction $a \rightarrow b + \phi$ can occur, then so can the reaction $\phi \rightarrow a + \bar{b}$. Consider Fig. 48, which depicts the creation of this pair from the neutral Higgs fields in the stationary Higgs sea. To

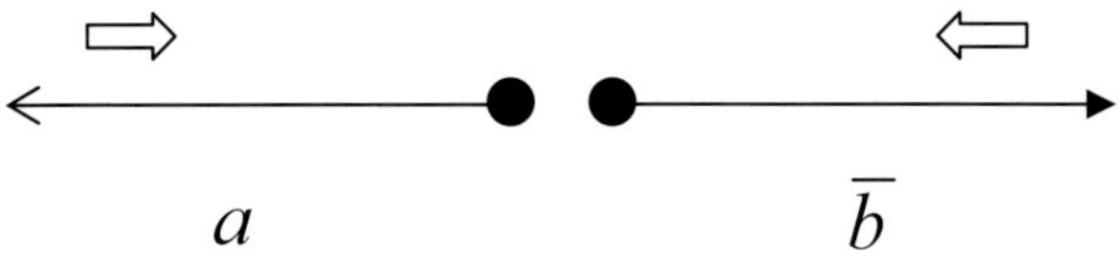

Figure 48: A pair of left-handed fermions a and $\bar{b}$ are created from the Higgs sea. The single arrows show their momenta and the double arrows their spins along the momentum direction. It is also possible to create a pair of right-handed fermions by reversing the orientation of both double arrows.

conserve momentum, the two produced particles a and $\bar{b}$ must go in opposite directions, as shown in Fig. 48. To conserve the total angular momentum component along the direction of motion, they must have the *same* helicity.[2] If a and b are both electrons, then the creation in Fig. 48 is allowed because it conserves momentum, total angular momentum, and electric charge, and because both electrons and positrons have right-handed as well as left-handed helicities. As remarked above, the square of the electron mass is proportional to the probability of this reaction taking place. Similarly, a mass can be generated this way for all the other fermions in the SM, except the neutrinos.

The problem with neutrinos is that the neutrino is always left-handed and the anti-neutrino is always right-handed, but Fig. 48 requires both of them to be left-handed (or right-handed). Hence, this process cannot occur and neutrinos remain massless in the SM.

Neutrinos

I just explained why neutrinos have to be massless in the SM. However, 'neutrino oscillation' experiments in recent years show us that at least two of the three neutrinos must have a tiny mass. According to Fig. 48, this can happen only if a left-handed anti-neutrino exists. We are therefore left with two possibilities. Either both a and $\overline{b}$ in Fig. 48 are the left-handed neutrinos, or else a hitherto unsuspected left-handed anti-neutrino $\overline{N}$ exists. Both possibilities have non-trivial consequences.

In the first scenario, the number of neutrinos is no longer conserved, because a pair of neutrinos can emerge from the Higgs sea which has no leptonic number. Since the neutrino carries no electric charge, the only thing that distinguishes a neutrino from an anti-neutrino is the leptonic number (neutrino has $L = 1$ and anti-neutrino has $L = -1$). Now that the leptonic number is no longer conserved, there is nothing meaningful that can distinguish the two anymore so we may simply regard the neutrino and the anti-neutrino to be one and the same. A fermion which is its own anti-particle is called a *Majorana fermion*, in honor of Ettore Majorana, who first proposed this possibility. Since an anti-particle has the opposite electric charge as the particle, only electrically neutral fermions may be Majorana fermions. Among the fundamental fermions, a neutrino is therefore the only candidate. The usual kind of fermions (all the quarks and all the charged leptons) whose anti-particles are distinct from the particles are known as *Dirac fermions.*

If we consider the neutrino ν as a Majorana fermion so that it may have its own mass, then we should no longer refer it as the left-handed neutrino because it is now the embodiment of the left-handed neutrino and the right-handed anti-neutrino. Since it has a small mass, henceforth we will refer to it as the *light neutrino.*

Whether the neutrinos are indeed Majorana fermions or not can only be determined by experiments. The best confirmation of this possibility at present is to detect a *neutrinoless double beta decay.*

Beta decay changes a neutron into a proton and creates a pair of leptons. Occasionally, two neutrons inside a nucleus both undergo beta decay in a process known as a double beta decay.

Since this is a doubly weak process, double beta decay is very rare but it has already been detected. Now, if the neutrino is its own anti-particle, then there is a possibility that the two neutrinos in double beta decay will annihilate each other, leaving behind a final product without any neutrino at the end. The detection of these neutrinoless double beta decay processes is a proof that neutrinos are their own anti-particles. Many experiments have been set up to detect such decays, but unfortunately the decay rate is expected to be very low and to date it has not yet been detected.

Another possibility is the existence of a separate and new right-handed neutrino N, and a left-handed anti-neutrino $\bar{N}$. Since these particles have not been seen, they are presumably heavier than our present capability to produce or to detect.

In the case of charged fermions, what appears in Fig. 48 is a particle–anti-particle pair with the same helicity, e.g. a left-handed electron and a left-handed positron, or a right-handed electron and a right-handed positron. Figure 48 is symmetric between the particle and the anti-particle so both of them have the same mass, as we already know. In the case of neutrinos, we can have a pair consisting of a left-handed ν and a left-handed $\bar{N}$, or a pair consisting of a right-handed N and a right-handed $\bar{\nu}$. Like the charged fermions, the ν mass m will now be identical to the N mass. This is no good because the mass of ν is known to be tiny but the mass of N must be huge in order not to have been seen. The only way for N to have a much heavier mass M is for it to be a Majorana fermion and have in addition a coupling in which both a and b in Fig. 48 are equal to N. The mass m is usually called a *Dirac mass,* and the mass M is called a *Majorana mass* (for N). Since N is involved in both kinds of couplings, a shift in masses can be shown to occur. Mathematics shows that the resulting $N = \bar{N}$ mass is still very close to M, but the resulting $\nu = \bar{\nu}$ mass has undergone a huge shift and is now given by $m_\nu = m^2/M$. This formula is known as the *see-saw* mass formula, because a high M implies a low m_ν and *vice versa*, just like a see-saw. Henceforth, we will refer to the Majorana neutrino N as the heavy neutrino.

We do not know the magnitudes of m and M, but we do know that M is much bigger than m. Since the reaction in Fig. 48 generating the Dirac mass m is similar to the reaction generating all the charged fermion masses, we suspect m is about the same size as the charged fermion masses. This actually says very little because the charged fermion mass ranges widely, from 0.51 MeV for the electron to over 170 GeV for the t quark (the third generation equivalent of the u quark). We also know that the neutrino mass m_ν is quite small, possibly less than 0.1 eV = 10^{-10} GeV. To have an estimate of what M could be, suppose we take m to be 1 GeV, then $M = 10^{10}$ GeV, certainly not a mass that we can hope to detect directly in the foreseeable future.

To summarize, the detection of neutrino mass in 'neutrino oscillation' experiments can be explained in two ways. Either the neutrino ν is a Majorana neutrino, or there exists a new right-handed neutrino N with a heavy Majorana mass M. The first alternative does not explain why the neutrino mass is so much smaller than the charged fermion masses. Compared to even the smallest of the charged fermion masses, of 0.51 MeV for the electron, the neutrino mass of 0.1 eV is five million times smaller. The second alternative can explain (through the see-saw mechanism) why the neutrino mass is so small. For that reason, we shall assume from now on that the second mechanism prevails. The leptogenesis mechanism for generating extra matter depends completely on that plausible assumption.

By the way, as mentioned before, the Majorana nature of the light neutrino ν inherent in the first mechanism can be confirmed by detecting neutrinoless double beta decay. However, the mere detection of this process cannot distinguish these two mass generating mechanisms, because the second mechanism also gives rise to neutrinoless double beta decays for the following reason.

The presence of the Dirac mass reactions $\phi \to \bar{\nu} + N$ and $\phi \to \nu + \bar{N}$ implies the presence of the decay reactions $N \to \nu + \phi$ and $\bar{N} \to \bar{\nu} + \phi$, and more generally $N \to \ell + \phi$ and $\bar{N} \to \bar{\ell} + \phi$ where ℓ is any lepton, charged or not. If it is charged, then of course ϕ has to be charged to uphold charge conservation. Now if N is Majorana, being its own anti-particle, then two N's can annihilate each other, so the number of N particles can change by 2. If that is the case, then the number of ν and more generally the number of leptons can change by 2. This gives rise to two important consequences. Firstly, a light neutrino can annihilate with another light neutrino to produce neutrinoless double beta decay. Secondly, the leptonic number L of the observed leptons in the SM need not to be conserved either. The reason why such violations have not been observed in the laboratory is because there are no *N's* present to make that possible.

As will be explained in the last section, the leptogenesis mechanism relies on the decay of N to generate extra anti-leptons, and the sphaleron mechanism to convert some excess anti-leptons into nucleons. We shall next describe what the sphaleron mechanism is.

Sphaleron

A sphaleron is a complicated static configuration of the electroweak bosons that requires a certain amount of energy to build. It is not a particle because it does not move, but like the mass of all particles in the SM, the energy required to build the sphaleron goes to zero in the symmetric phase above the electroweak transition temperature. The existence of such a configuration can be shown in the SM theory.

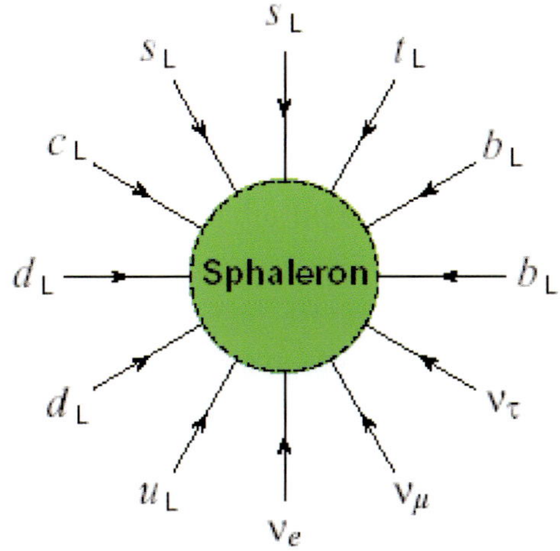

Figure 49: A sphaleron process. See Fig. 47 for the names of the particles.

The job of a sphaleron is to catalyze the production or the disappearance of 12 left-handed fermions together. These 12 are made up of three left-handed quarks of three different colors and one left-handed lepton taken from each of the three generations, in any combination provided their total charge is zero to uphold charge conservation. Figure 49 shows the disappearance process, in which the subscript L indicates that these fermions are left-handed.

Since three quarks of three different colors make up a baryon, this mechanism catalyzes the creation or the disappearance of three baryons and three leptons. As a result it changes the baryonic number B by three units, and the leptonic number L also by three units. It changes $B + L$ by six units, but it leaves $B - L$ unchanged.

Like all particle reactions, we may move some of these fermions in the final state to the initial state, and *vice versa*, provided fermions are changed to anti-fermions when they are so moved.

For example, it can convert three anti-leptons from three different generations into three baryons. Baryons are by definition particles with three quarks. If all three quarks come from the first generation, then the baryon is a nucleon. If not, these baryons are heavier but unstable, and each of them will eventually decay into a nucleon plus something else, so the generation of baryons automatically leads to the generation of nucleons.

The amount of energy required to build a sphaleron below the electroweak phase transition temperature is about 10 TeV (1 TeV = 10^{12} eV). Since the phase transition temperature is about 100 GeV, that kind of energy is never available thermally, so the reactions catalyzed by the sphaleron never happen in practice below that temperature.

Above the phase transition temperature, no energy is needed to build it, so these catalytic reactions proceed very rapidly. We might therefore think that sphalerons could be used to produce the baryon and lepton excess, but quite the contrary: they actually tend to wipe out any excess that may have been built up by other means. For example, if there is a mechanism to produce an extra $B = L \neq 0$ above the phase transition temperature, eventually both B and L will be reduced to zero by these fast catalytic reactions. For a matter generation mechanism to be useful it must also have $B - L \neq 0$, for then it cannot be wiped out completely by these catalytic reactions which conserve $B - L$. In fact, calculations show that under thermal equilibrium, about 1/3 of the $B - L$ amount will turn into the B number, and about –2/3 of that amount will turn into the L number. If $B - L = 0$, then there will be no net B or L left at the end. In the leptogenesis scenario described below, an excess of anti-leptons is obtained from the N decay. The sphaleron mechanism then converts one third of the excess anti-leptons into nucleons.

This concludes our discussion on particle physics. In the next section we will discuss the meaning of the three Sakharov processes, why they are necessary, and why it is so difficult to satisfy them.

The Sakharov Conditions

To recapitulate, the three Sakharov conditions that must be satisfied in order for a net leptonic number L (or a net baryonic number B) to be generated are:

(i) Baryonic or leptonic number conservation must be violated.
(ii) C and CP invariance must be violated.
(iii) Thermal equilibrium must be violated.

Let me discuss these conditions one by one.

The First Condition

This condition is obviously necessary, for otherwise if we start with $L = 0$ we will end up with $L = 0$. It is hard to satisfy because we know of only two mechanisms that cause L conservation to be violated. The first is the sphaleron mechanism, and the second is the presence of Majorana neutrinos. Both are very feeble at the present temperature so if L violation happens it presumably happened when the universe was hot. The leptogenesis mechanism actually makes use of both of them. It relies on the decay of the heavy Majorana neutrinos N to generate extra anti-leptons, and it relies on the sphaleron mechanism to convert some of these anti-leptons into baryons.

The Second Condition

This condition is necessary for the following reason. Let the leptonic number *density* at time t be denoted by $L(x,y,z,t)$. Then, the total leptonic number at a time t is given by integrating $L(x,y,z,t)$ over all spatial positions (x,y,z). Under the C operation, particles are changed to anti-particles at every spacetime point, and hence the density $L(x,y,z,t)$ becomes $-L(x,y,z,t)$. Integrating over all spatial positions, we conclude that $L = -L$; hence, $L = 0$ at all time. This shows that without C violation we can never get a non-zero leptonic number L.

A similar argument can be made to show that if the production process is invariant under CP, then the total leptonic number L must again be zero. This is so because under such an operation, $L(x,y,z,t)$ changes to $-L(-x,-y,-z,t)$. After integrating over the spatial coordinates (x,y,z), we obtain once again $L = -L$; hence, $L = 0$ at every t. This explains why the second Sakharov condition is necessary.

Same arguments also apply to the baryonic number B.

Why is this condition so difficult to satisfy? This is because in the SM, only very few weak-interaction reactions violate C and CP, and the probability of them occurring is rather low.

In leptogenesis, the presence of this new heavy Majorana neutrino N gives rise to other possibilities, and it is not hard to arrange its decay to violate both C and CP.

The Third Condition

The necessity of the third Sakharov condition can be proven but we do not possess the technical apparatus to do so here. Roughly speaking, what happens is the following. Under CPT, which is always an invariant, $L(x,y,z,t)$ becomes $-L(-x,-y,-z,-t)$. After spatial integration, the CPT theorem states that L at time t is equal to $-L$ at time $-t$. This by itself does not mean that L has to be zero either at time t or at time $-t$. However, if we assume thermal equilibrium, then everything is in a steady state, so time does not matter. In that case we have $L = -L$ at all time and hence $L = 0$.

It is difficult to satisfy this condition because it is difficult for particle reactions to get out of thermal equilibrium. There are two mechanisms we know of that accomplish the requirement. One is for the reaction to occur much more slowly than the expansion rate of the universe, and the other is for the reaction to occur during a *first order phase transition.* I will explain them separately below.

Thermal equilibrium is established when particles are allowed to undergo many collisions and reactions. If there is not enough time to do it, thermal equilibrium may not be achieved. This is the case when the universe expands faster than the rate of reaction or decay, because then the condition of the universe would have changed before thermal equilibrium can be established. As we shall see later, leptogenesis relies on this mechanism to achieve thermal non-equilibrium.

As for first order phase transition, a typical example is water at 100°C, under one atmospheric pressure. At that temperature, the liquid phase and the vapor phase coexist, so the system consists of two distinct densities, that of liquid water and that of steam vapor. It is not homogeneous and uniform, so it is not in thermal equilibrium. A change of this kind where two distinct phases can coexist is called a first order phase transition.

If we increase the atmospheric pressure, the boiling point gets higher, but since vapor is compressible and liquid water is not, their densities at the boiling will also become closer together. At a sufficiently high pressure, their densities become equal, the two phases are not so distinct and the phase transition becomes *second order.* Sakharov's third condition will no longer be satisfied when that happens.

There are models in which extra baryons and leptons are generated at the electroweak phase transition. However, in order for the phase transition to be strongly first order, so that enough deviation from thermal equilibrium is present to produce enough matter over anti-matter, the presence of new particles as well as a delicate adjustment of parameters are needed. We shall not discuss this kind of model here.

In conclusion, we see that the leptogenesis mechanism is capable of satisfying all three Sakharov conditions. In the next section, we will discuss how it works in more detail.

Leptogenesis

A Summary of the Mechanism

According to this theory, the emergence of matter over anti-matter can be traced back to the decay in the early universe of the heavy neutrinos *N*, which are their own anti-particles.

N could be produced during reheating after inflation. Their subsequent decay can be arranged to generate more anti-leptons than leptons, then the sphaleron mechanism converts one third of the excess anti-leptons into baryons. Like nucleons, all baryons are made up of three quarks. Since there are six different kinds of quarks, baryons are more general than nucleons which are always made up of the *u* and the *d* quarks. However, all the other baryons are unstable, and eventually they will all decay into nucleons plus something else. Thus, the excess baryons produced are the same as the excess nucleons produced.

For this mechanism to be viable, the masses of the ordinary neutrinos ν and $\overline{\nu}$ must be rather small, consistent with what we know them to be. This agreement gives us confidence about the correctness of the theory.

In order to produce enough nucleons to explain the measured number $\eta \cong 6 \times 10^{-10}$, the heavy neutrino *N* must have a mass not less than 10^9 GeV. This sets a lower bound on the allowed reheating temperatures, a bound which we have already used before.

Details of the Mechanism

There may be more than one kind of right-handed neutrino N, and if that is so, we assume one of them, N_1, to have a mass M_1 much lower than the others. As will be explained below, in that case only the decay of the lightest counts. So, for all practical purposes, we can forget about all of them except N_1.

The reason is because the number of heavy neutrinos decreases slowly with the temperature T until the temperature reaches down to its mass. Below that point the number drops rapidly to zero because there is no longer enough thermal energy to recreate the heavy neutrino once it has decayed away. Since the final decay products are the same for all the different kinds of heavy neutrinos, as long as the temperature is larger than M_1, the decay products can always recombine to form N_1. This is why the final number of decay products is determined at the temperature $T = M_1$ by the decay of N_1.

Sakharov's first condition is obeyed because N is a Majorana neutrino so leptonic number is no longer conserved. We can and will arrange the decay $N \to \ell + \phi$ to violate C and CP so that Sakharov's second condition is met. To fulfill Sakharov's third condition, this decay must be out of thermal equilibrium, and that requires the decay rate to be smaller than the Hubble expansion rate H. In the radiation era when this takes place, H is equal[3] to αT^2, and hence αM_1^2, where α is some known constant. On dimensional grounds, the decay rate is proportional to $s_1 M_1$,[4] where s_1 is the coupling of N_1 to ℓ and ϕ. Sakharov's third condition therefore requires $s_1 M_1 < \alpha M_1^2$, which implies $s_1 < \alpha M_1$.

From the see-saw mechanism, we know that the mass ν_1 of the lightest neutrino is equal to s_1' / M_1, with a s_1' that can be shown to be less than s_1. But according to the inequality in the last paragraph, s_1 is itself less than αM_1. Thus, the mass of the lightest neutrino ν_1 must be less than α, a calculable number which turns out to be about 5 meV (milli-electronvolt). We shall come back to the significance of this requirement later.

With the Sakharov conditions satisfied, it is possible to obtain an excess of anti-leptons over leptons, and via the sphaleron mechanism, an excess of nucleons over anti-nucleons. We shall now outline why in order to obtain enough nucleon excess to agree with the measured value of η, the nucleon-to-photon ratio, the mass M_1 of the lightest heavy neutrino must be larger than 10^9 GeV.

There are four factors contributing to the size of η. The first is how many N_1 there are. Since we assume they were produced in thermal equilibrium, that number density is known[3] and depends only on the temperature $T = M_1$. The second factor is related to the second Sakharov condition, and it depends on how much CP violation there is. This is measured by the factor ε_1, which is defined to be the difference of the decay rates of $N \rightarrow \ell + \phi$ and its CP-conjugated process $\bar{N} \rightarrow \bar{\ell} + \bar{\phi}$, divided by the sum. It is non-zero as long as the coupling constant is complex, in which case CP is violated. Clearly, by definition the absolute magnitude of ε_1 cannot be larger than 1, but it can be shown that it cannot be larger than some known number times M_1. The third factor is related to the third Sakharov condition, and it measures how far from thermal equilibrium the decay process is. This *wash-out* factor κ is zero if everything is in thermal equilibrium. It requires numerical simulation to compute but typically it is not larger than about 0.1, meaning that the maximum efficiency for generating matter is not more than 10% or so in the mass range we are interested in. These three factors determine how many extra leptons are produced. The fourth factor is the efficiency of the sphaleron mechanism turning anti-leptons into baryons. As discussed before, this factor is known to be about a third.

Now the first two factors increase with M_1. Hence, to get a large enough value of η, the mass of the lightest heavy Majorana fermion N_1 cannot be too small. If we put in the numbers, this lower bound on M_1 comes out to be about 10^9 GeV, as stated before. This also means that the reheating temperature at the end of inflation must be at least that much.

These estimates on the mass upper bound of ν_1 and the mass lower bound of N_1 assume the decay process gets out of thermal equilibrium suddenly at temperature $T = M_1$. In reality, everything proceeds more gradually, and it requires detailed numerical calculations to get the precise numbers. The conclusions of such calculations are that

(i) the mass of all the neutrinos ν must be less than 0.13 eV;
(ii) leptogenesis works best when the neutrino masses are in the window between 1 meV and 0.1 eV;
(iii) the reheating temperature must be larger than 10^9 GeV.

What is very interesting about the first two conclusions is that this range of light neutrino masses is consistent with measurements of neutrino masses by particle physics methods. This provides great encouragement to the leptogenesis process as the correct mechanism for producing the excess nucleons in our universe.

19

Syntheses of Chemical Elements

Figure 50 shows the relative abundance of chemical elements in the universe and in the crust of the earth, plotted in a logarithmic scale. In that scale, one unit on the vertical axis corresponds to a factor ten difference in the quantity.

The two curves more or less follow each other, with some notable differences. There is a great depletion of noble gases (Helium, Neon, Argon, etc.) in the crust, presumably because they are inert and cannot be retained in the crust by forming compounds with other elements. There is also a shortage of hydrogen (H) and helium (He) in the crust, and as we know, also in the atmosphere, because the small gravity of the earth cannot hold onto these light gases. It is different in big gaseous planets like Jupiter and Saturn, where gravity is so strong that plenty of H and He are found in their atmospheres.

What we would like to discuss in this chapter is where these elements in the cosmos come from. We will concentrate on the origin of their nuclei, not the neutral atoms *per se*, because the latter are ionized either before decoupling or in the vicinity of a star.

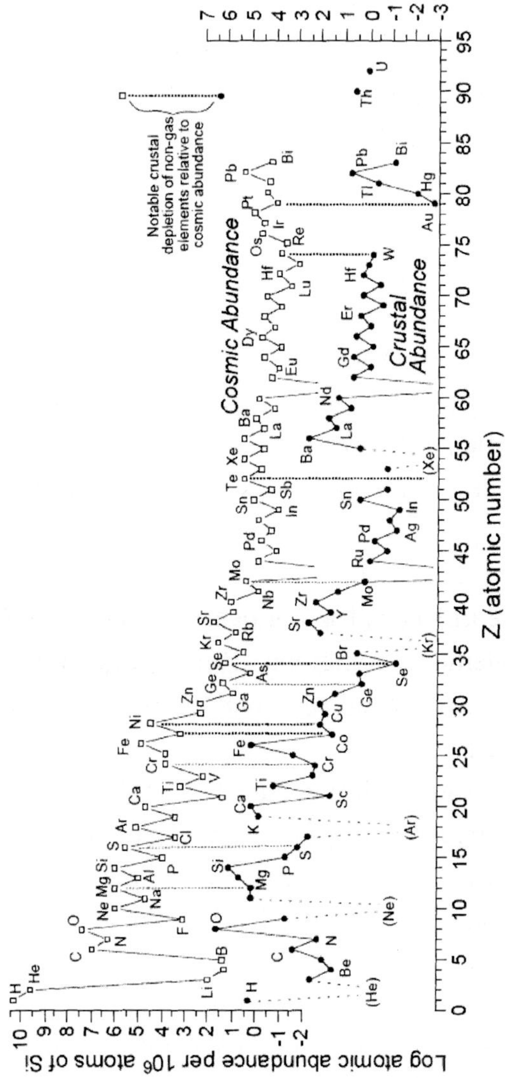

Figure 50: The cosmic abundance (upper curve) and the earth's crustal abundance of chemical elements. The former is taken from L. H. Ahrens, *Distribution of the Elements in Our Planet*, McGraw Hill (1965). The latter is taken from K. B. Krauskopf, *Introduction to Geochemistry* (2nd edn.), McGraw Hill (1979).

To find out where the nuclei in the universe come from, let us look at the abundance curve for some hints. The most noticable feature is that the abundance zigzags, showing elements with even atomic numbers (namely, even number of protons in their nuclei) more abundant than their neighboring elements with odd atomic numbers. This feature is due to the presence of a *pairing force* between nucleons, making nuclei with even atomic numbers more tightly bound and more stable than those with odd atomic numbers.

The other general feature is that the abundance as a whole decreases with increasing atomic numbers. This suggests that the heavier elements are somehow assembled from the lighter elements.

The most remarkable feature, though it may not be immediately obvious in this logarithmic plot, is that most of the universe is made up of hydrogen and helium. In spite of their lightness, they comprise about 74% and 24%, respectively, of the *weight* of the whole universe, leaving behind only about 2% for all the other heavier elements combined. This suggests that somehow hydrogen and helium are singled out and produced in a different manner than the rest.

The hydrogen nucleus is just the proton, and we already found out in the last chapter where protons and neutrons came from, possibly through the leptogenesis mechanism. At the time of their production, the temperature is likely to be above 10^9 GeV, so none of the other nuclei can possibly exist because their binding energies are much lower, in the neighborhood of MeV.

Some helium is produced in the stars, but most was already formed in the early universe at a temperature of a MeV or so, in a process known as *Big Bang Nucleosynthesis* (BBN). This occurred in the first few minutes in the life of the universe; other than helium only trace amounts of very few other nuclei were formed at that time.

All other nuclei up to iron (Fe) were produced in the interior of stars. They are the ashes left over from star burning.

Nuclei heavier than iron are believed to be formed during supernova explosions. The fact that our body and the earth contain all these elements shows that we are not only made from ashes of stars, but also remnants of supernova explosions.

I will elaborate a bit on each of these three mechanisms.

Big Bang Nucleosynthesis (BBN)

After annihilation of all the anti-nucleons, the leftover protons and neutrons maintain a thermal equilibrium through the weak processes $n \leftrightarrow p + e^- + \overline{\nu}_e$, $\nu_e + n \leftrightarrow p + e^-$, and $e^+ + n \leftrightarrow p + \overline{\nu}_e$. Since the neutron is 1.293 MeV heavier than the proton, it is energetically more demanding to change a proton into a neutron than the other way around; hence, there are always a little fewer neutrons than protons in the equilibrium mixture. The difference is negligible at high temperatures, but it becomes quite noticeable at a temperature of about 1 MeV.

Several things happened in succession then. Just before reaching that temperature, neutrinos froze out of thermal equilibrium (see Chap. 18 for a discussion of freeze-out) when their weak reaction rates dropped below the expansion rate of the universe. This also put the neutron-to-proton ratio maintained by the processes above out of thermal equilibrium and approximately frozen at a ratio 1/6. From there on, this number decayed only slowly due to the slow decay of neutrons.

Later on, at about 1/3 the electron mass of $m_e = 0.5$ MeV, positrons disappeared from the universe through annihilation with electrons into photons. The energy so released heated up the universe by a factor $(11/4)^{1/3} = 1.40$.[1] This heating applied to everything except the neutrinos and anti-neutrinos because they were already decoupled from the rest of the universe. As the universe continued to expand, both temperatures decreased like $1/a$, keeping this ratio constant. Thus, the measured photon temperature of 2.725 K today predicts the temperature of the cosmic background neutrinos to be 2.725/1.4 = 1.95 K.

Since neither two protons nor two neutrons can form a stable nucleus, the first complex nucleus to appear by two-body collisions was the deuteron, which was made up of one proton and one neutron with a binding energy of 2.22 MeV. Deuterons began to appear after positron annihilation took place.

One curious feature in the narration above is that all the events seem to be happening at too low temperatures. Why did electron–positron annihilation take place at 1/3 of m_e rather than m_e? Why did the deuterons begin to appear only after positron annihilation and not around its binding energy of 2.22 MeV?

The reason is the smallness of the number η.

When a neutron and a proton snap together to form a deuteron, the released binding energy is carried away by a photon. Conversely, under thermal equilibrium, a photon with such an amount of energy can hit a deuteron and photo-dissociate it back into a proton and a neutron, thus breaking up the deuteron again. Normally, this photo-disassociation would stop at a temperature below the binding energy of 2.22 MeV, but with $\eta \cong 6 \times 10^{-10}$ the situation is not normal. Remember from Fig. 34 that photons carry a range of wavelengths and hence a range of energies, though the probability of a photon carrying an energy much larger than the temperature is small. However, this number η tells us that there are more than a billion photons to each nucleon present; hence, a deuteron can still be photo-disintegrated at a low temperature by one of these photons provided the probability of having a

2.22 MeV photon at that temperature is more than one in a billion. This is why the formation temperature is so low. A similar reasoning explains why positrons disappeared at 1/3 of m_e rather than m_e.

Once deuterons were formed, two of them could snap together to form a helium nucleus comprising of two protons and two neutrons. Since each deuteron carries one unit of positive charge, to fuse them together we must overcome their electric repulsion, and that requires a temperature above 35 keV (1 keV = 10^3 eV). That is not a problem at a temperature of about 0.1 MeV = 100 keV when this took place.

The helium nucleus is very stable and has a large binding energy of 28.3 MeV. As a result it converts most of the deuterons into helium.

By that time, at a temperature of about 0.1 MeV, the neutron–proton ratio had decreased to about 1/7, namely, for every two neutrons present there were 14 protons around. The two neutrons are taken up together with two protons to form a helium nucleus, leaving 12 protons behind as hydrogen. Hence, the helium/hydrogen ratio in the universe *by weight* is about 4/12 = 1/3. In other words, approximately 25% of the weight of ordinary matter is carried by helium, and 75% is carried by hydrogen.

Deuterons can also fuse with hydrogen to form He^3, an isotope of helium with two protons and a neutron. The number on the upper right corner indicates the total number of nucleons; thus, the usual helium discussed above is He^4. However, since the binding energy of He^3 is only 7.72 MeV, its abundance is far less than the abundance of He^4.

In principle, more complex nuclei could be formed from other reaction processes, but to fuse into higher elements with more protons, more electric repulsions must be overcome which require even higher temperatures. By this time, the temperature was getting pretty low, and getting lower all the time as the universe expanded. There is no stable nucleus with five nucleons to make it easier to produce, so the only other element produced by BBN was Li^7 (lithium), and that only in a trace amount because of the temperature of the universe by that time.

Looking back, the reason why heavier elements are not produced in BBN can be traced back to the smallness of η and the smallness of the deuteron binding energy, both compelling deuterons to be formed at a relatively low temperature. Without first forming deuterons, it is hard to produce the heavier elements, but with the smallness of η and the deuteron binding energy, after they are formed the temperature has gone too low to allow heavier elements to be fused. This bottleneck occurring from the late formation of deuteron is sometimes known as the *deuterium bottleneck*.

Deuterons have a small binding energy, so they are easily destroyed in the stars. Thus, whatever amount of deuterons observed today cannot be bigger than the actual amount produced by BBN. The other BBN elements can both be created and destroyed in the stars, so it is much harder to figure out for them the exact relationship between the amount observed and the amount produced in BBN. To avoid these complications, it is best to measure the BBN abundance in a region where star activities can be minimized.

The measured amount of deuteron corresponds to a value $\eta = 5.95 \pm^{0.36}_{0.39}$, which is consistent with the abundance of the other BBN elements like He^3, He^4, and Li^7.

This value is also in agreement with that obtained from the CMB. This is very interesting because BBN occurs at a temperature of 0.1 MeV, and CMB emerges at a temperature of ¼ eV. The two events are some 400,000 years apart, yet they both yield the same value of η giving powerful support for the correctness of our cosmology theory. Note that both nucleon and photon number densities are proportional to $1/a^3$ so their ratio should remain the same at all times.

Before ending this topic, let me remark that all this would be completed about half an hour after the Big Bang. Please see footnote [2] to find out how to relate temperature and time.

Fusion in Stars

In Chap. 12, we learned that the sun's energy was derived from fusing four protons into a helium nucleus, plus two positrons and two neutrinos. This fusion takes place through several steps with intermediate products, but the details need not bother us. Because of the electric repulsion between protons, fusion can take place only at a high temperature, in this case about 15 million degrees. This high temperature is in turn derived from the energy released in gravitational contraction.

At a higher temperature, there is a more complicated pathway to burn hydrogen into helium, some relying on the presence of a small amount of carbon or nitrogen acting as catalysts, others utilizing the help of the BBN helium already present.

When hydrogen is exhausted at the center of the star, burning may proceed to the outer layers, leaving only helium ash at the center core. With the fire extinguished, there is no longer anything to stop gravitational contraction at the center. If the star is large enough to allow a sufficient amount of gravitational energy to be released during this contraction, the temperature at the center may rise to 100 million degrees, in which case helium burning can start. Higher temperature is needed to fuse nuclei of larger atomic numbers because more electric repulsion has to be overcome.

Two He^4 nuclei can fuse into a Be^8 (beryllium, with four *p* and four *n*), but that is unstable. Three He^4 can fuse into a C^{12} (carbon, with six *p* and six *n*), four He^4 can fuse into a O^{16} (oxygen, with eight *p* and eight *n*), and five He^4 can fuse into a Ne^{20} (neon, with ten *p* and ten *n*). These elements are all tightly bound because helium is and they are made out of several helium nuclei. You can see in Fig. 48 that these elements are some of the most abundant ones after H and He^4. Other processes can occur to generate other elements but we will not talk about them in detail.

Now the story more or less repeats itself. If the star is large and the carbon ash at the core can rise to a temperature of 600 million degrees, carbon may fuse with itself to produce Mg^{24} (magnesium, with twelve *p* and twelve *n*), or carbon may fuse with a helium to get O^{16}. When the temperature reaches 1,200 million degrees, two neon can fuse into a Mg^{24} and a O^{16}. When the temperature reaches 1,500 million degrees, two oxygen can fuse into helium and a Si^{28} (silicon, with fourteen *p* and fourteen *n*). When the temperature reaches 2,700 million degrees, two silicon can fuse into a Ni^{56} (nickel, with twenty-eight *p* and twenty-eight *n*). Now Ni^{56} is not stable: it will decay rapidly into Fe^{56} (iron, with twenty-six *p* and thirty *n*) by turning two protons into two neutrons (plus two positrons and two neutrinos). As before, other nuclear reactions can occur to produce other elements.

A star like the sun takes about 10 billion years to burn out its hydrogen, but the burning of higher elements in larger stars proceeds much faster. For 25 solar mass stars which can burn silicon, the silicon in its core will be all burned up in a day.

When iron is reached, no further burning can take place because iron has the largest nuclear binding energy, as shown in Fig. 51. Burning is allowed to release energy when a tighter binding is obtained by fusing two nuclei. When the tightest

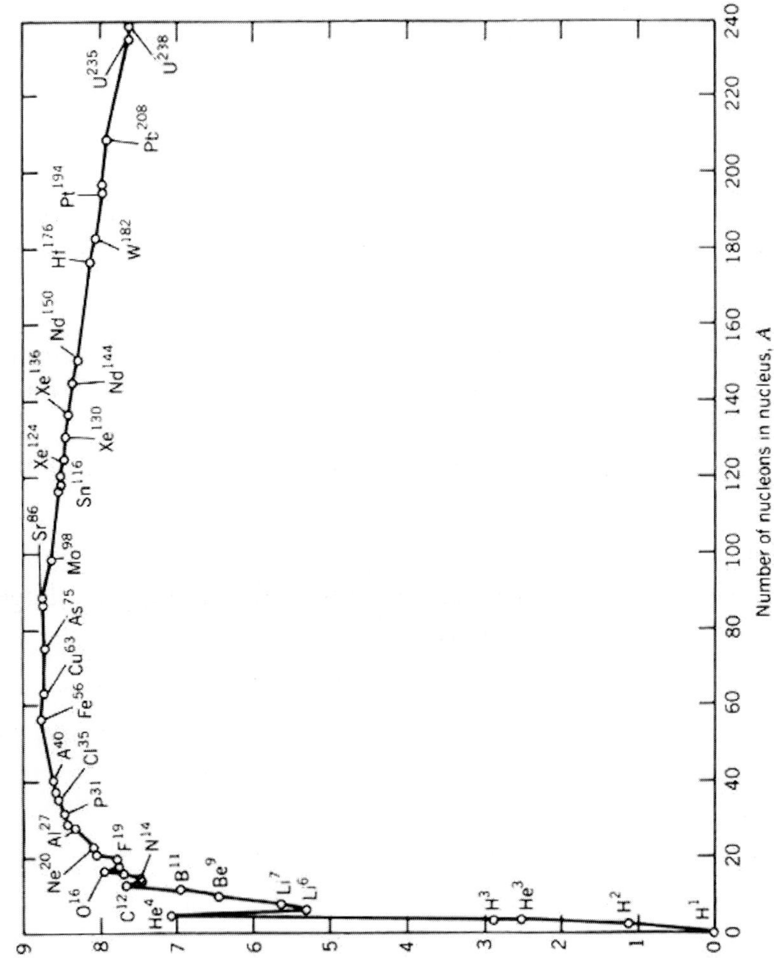

Figure 51: Nuclear binding energy per nucleon as a function of the number of nucleons in the nucleus.

binding is reached around iron, no further energy can be derived by fusing them so these elements can no longer be burned.

Supernova Explosion

To synthesize even heavier elements both energy and nucleons must be supplied. It is better to add neutrons than protons to avoid the problem of electric repulsion. The super neutron-rich nuclei thus created are unstable, but that does not matter because neutrons can beta decay into protons to produce stable elements heavier than iron.

The process which allows nuclei to assemble lots of neutrons quickly is known as the *r-process,* with '*r*' standing for 'rapid.' We will not discuss the details of the *r*-process here as some of them are still being worked out.

To generate heavier elements this way, we need to have a source which can supply both energy and neutrons. The prime candidate for that is a *supernova*. I will spend the rest of this section to sketch what a supernova is, and why it is able to supply a large amount of energy and neutrons.

When a star burns, it generates a pressure to halt the gravitational contraction of the star. After the fuel is exhausted, the pressure is gone so gravitational collapse resumes again. Unless there is something to stop it, the size of the dead star will keep on shrinking, eventually becoming so small and the gravitational field in its vicinity so strong that nothing in its vicinity can ever escape from its gravitational hold, not even light.[3] Such a dead star appears to us to be black and is thus called a *black hole.* This would be the fate of a dead star when it is very massive.

For a small star with a mass less than 1.4 solar masses, Subrahmanyan Chandrasekhar discovered that because of the Pauli exclusion principle,[4] shrinking would stop when the electrons touched one another. This limiting mass is known as a *Chandrasekhar limit*, and the resulting dead star is known as a *white dwarf*.

For stars not much heavier than the Chandrasekhar limit of 1.4 solar masses, a similar reasoning shows that neutrons in the star can uphold a star when electrons fail. Since electrons alone cannot stop the gravitational collapse, electrons in atoms will all be squeezed into the nuclei as the star contracts. They are then captured by the protons and turned into neutrons after emitting neutrinos. Now the star has more neutrons than it had electrons, because on top of the neutrons converted from protons and electrons, there are also the original neutrons in the nuclei. With more neutrons the star is able to resist the gravitational crunch of a larger star. The resulting dead star is called a *neutron star*, because it contains mostly neutrons.

As with electrons, this will not take effect until the neutrons are touching one another. That means the density of the neutron star is comparable to the density of a nucleus. This also means that all the mass of one and half of the sun is concentrated in a radius only 10 to 20 kilometers across!

The star has a long way to fall from its original size down to 10 or 20 kilometers before the contraction stops. When it finally reaches the incompressible neutron core, it cannot fall any further, and that causes a big bounce. This is like dropping an elastic ball from the top of a skyscraper: when it hits the ground, it cannot fall any further so it bounces back up. The bounce from the falling star generates a big explosion accompanied by a big shock wave. In this environment lots of energy is available, and lots of neutrons can be found from the neutron core to synthesize heavy elements via the *r*-process. To be sure, it is not at all trivial to work out all the details about how the energy and the neutrons are being transferred, but it is commonly believed that this is how heavy elements are built up.

This big explosive event emits so much energy that it suddenly lights up the dead star as if a new star was born: that is why it is called a nova, meaning a new star. This kind of explosion is so huge that the light output can often equal the light output of a whole galaxy, at least for a little while — hence the name *super*nova.

The first supernova recorded in history seems to be the one discovered by the Chinese in 185 CE, but the most famous one is probably the one seen by the Chinese and the Arabs in 1054 CE. It is situated in the constellation Taurus, located 6,300 light years from us, whose remnant seen below in Fig. 52 is called the *Crab Nebula*. The nebula now has a diameter of 11 light years, and the neutron star located at its center has also been detected.

Figure 52: The Crab Nebula, the remnants of a supernova exploded in 1054 CE.

20

Epilogue

We have come to the end of our journey, in an exploration to find out how the vastness and the richness of our present universe could have emerged from a tiny one with no matter and very little energy.

We have found that the chemical elements came from the fusion of protons and neutrons. Light nuclei like deuteron, helium, and a small amount of lithium were produced very early in the universe, but the other ones were assembled much later in stars and in supernovae.

We have discussed the leptogenesis mechanism for generating a net amount of matter in the universe. Neutrons and protons came from anti-leptons, and the excess anti-leptons came from the decay of heavy Majorana neutrinos, possibly produced during reheating.

We have learned that reheating derived its energy from the decay of the false vacuum at the end of inflation, and the explosion in the Big Bang extracted its energy also from the inflation. Inflation itself was driven by the constant energy density in the false vacuum, so everything boiled down to the presence of the false vacuum. At the moment this is the frontier of our knowledge; we neither know what preceded it nor what caused the energy density to remain constant for such a long time.

From inflation on, the evolution of the universe was at every moment controlled by the Friedmann equation, which can be nicely summarized by the Tai Chi symbol in Fig. 41.

Now that we have concluded our journey which began in Chap. 1, it is time to wrap up and ask what we have found out about the connection between the emptiness in Zen and the emptiness in Cosmology, a question which sent us on our way in the first place. Other than a common name, emptiness, clearly there cannot be a direct connection between the two, because the former is an attitude and a philosophy of life while the latter is a science based on physical evidence and experimentation. Yet, there are some subtle connections and links. Zen strives to have a mind as vast and as empty as our universe, and cosmology is precisely the science of this vast and empty universe. Zen acknowledges the presence of stars and galaxies but does not consider them ruinous to the emptiness, and cosmology wants to trace back even the origin of these stars and galaxies. It finds them originating from almost nothing in the primordial universe, which is itself a very Zen-like answer. Yet, as argued in Chap. 16, to a non-Buddhist and a layman, maybe neither Zen nor the primordial universe is really empty, though technically they can both be taken to be.

Appendix A: Endnotes

These notes are organized according to chapters. A reference such as 7[1] refers to note [1] in Chapter 7.

Chapter 7

[1] Take a galaxy d Mpc from us. Its receding velocity is then $v = 72d$ km/s. When we reverse its motion by running time backwards, it will take a time of $t = d/v$ to reach us. Since 1 Mpc is 3.26 million light years, and one light year is 0.946×10^{13} km, we get $t = 4.28 \times 10^{17}$ s $= 1.36 \times 10^{10}$ years.

[2] See note 15[4].

Chapter 8

[1] Eratosthenes knew that on the summer solstice the sun appeared right overhead at noon in the town Syene in Egypt, because it is on the Tropic of Cancer. He measured the shadow cast by a pole in Alexandria at noon on the same day and found that the sun was 7.2° away in the southerly direction from overhead. This means that the distance between the two cities is 7.2/360 of the total circumference of the Earth. He

hired a person to pace out the actual distance between the two cities, which turned out to be about 800 km. In this way he came up with a circumference of the earth to be 39,690 km, which is only 1% short of the actual circumference of 40,008 km, a remarkable feat indeed more than two thousand years ago. Knowing the circumference one knows the radius of the earth, and hence its curvature.

[2] In the absence of any force, particles move in straight lines with constant velocities, namely, they move along geodesics of a flat space, and a flat spacetime.

That is no longer true if forces are present. However, if that force is gravity, and gravity alone, then the force is proportional to the mass of the particle (Chap. 6), and hence (according to Newton's Second Law of Mechanics) its acceleration and its trajectory are *independent* of its mass. This special feature of gravity enabled Einstein to envisage the trajectories to be geodesics of an underlying curved spacetime. However, for that to work, different trajectories must be intimately connected to enable them to weave into a single fabric of spacetime. Einstein found that he had to change the Newtonian law of gravitation a little bit in order for that to succeed. This is the general theory of relativity; the modified laws of gravity have since been proven experimentally to be correct. Note that once the spacetime is curved, light must follow the geodesics and be bent by the presence of a massive object as well.

Chapter 9

[1] This assertion is true as long as the recession velocity of a galaxy is proportional to its distance at all times, allowing the proportionality constant to change with time. In that case,

although the distance to every galaxy changes with time, the ratio of our distances to two galaxies remains the same at all times; hence, it does not matter what galaxy you use to measure the scale factor a.

[2] Consider an electromagnetic wave of wavelength λ and period T emitted from a distant galaxy. According to the discussion above Fig. 16, these two quantities are related by $\lambda = cT$. Suppose a photon emitted at time t arrives on earth at time t_0, and a photon emitted one period later at time $t' = t + T$ arrives on earth at time $t_0' = t_0 + T'$. For a static universe, the perceived period T' at reception is the same as the emitted period T, and the perceived wavelength λ' at reception is the same as the original wavelength λ. Since the universe is expanding and the galactic source is receding, it would take longer for the second photon to reach us so we expect $T' > T$ and hence $\lambda' > \lambda$, giving rise to a redshift. To compute what it is, let us use comoving distance because that distance is not affected by the expansion of the universe.

In time dt, a photon travels an actual distance $c\,dt$, and hence a comoving distance $dx = c\,dt/a(t)$. The comoving distance between earth and the galactic source is therefore equal to the integral $\int c\,dt/a(t)$. If we compute it from the first photon, the integration limits are between t and t_0. If we compute it from the second photon, the integration limits are between $t' = t + T$ and $t_0' = t_0 + T'$. Since the comoving distance computed from both photons must be the same, we see that this integral between the limits t and $t + T$ must be equal to the integral between t_0 and $t_0 + T'$. Since the universe has hardly expanded during the short time intervals T and T', these two integrals are $cT/a(t) = cT'/a(t_0)$, and hence $\lambda/a(t) = \lambda'/a(t_0)$. Redshift z is defined to be $(\lambda' - \lambda)/\lambda$; hence, $z + 1 = \lambda'/\lambda = a(t_0)/a(t) = 1/a(t)$.

[3] When z is small, according to the formula derived above, t must be very close to t_0, and $a(t)$ very close to 1. This allows various approximations to be made. The distance of the galaxy is now $d = c(t_0 - t)$, and its velocity is $v = [d - a(t)d]/[t_0 - t]$, because $a(t)d$ is the distance of the galaxy at an earlier time t. Hence, $v/c = 1 - a(t) \cong [1 - a(t)]/a(t) = z$.

Chapter 10

[1] P. Natarajan and V. Springel, *Astrophysical Journal Letters* **617** (2004) 13.

[2] D. Clowe *et al.*, http://arxiv.org/pdf/astro-ph/0608407.

Chapter 11

[1] Strictly speaking, the rest energies of the neutron, proton, electron, and anti-neutrino must also be taken into account in calculating the energy advantage or disadvantage for a beta decay.

[2] For a non-relativistic particle, its momentum $\vec{p} = m\vec{v}$ is defined to be its mass times its velocity. If $\vec{r}$ is the position of the particle, then its *orbital angular momentum* is defined to be $\vec{r} \times \vec{p}$. Its total angular momentum is its spin plus its orbital angular momentum. Since $(\vec{r} \times \vec{p}) \bullet \vec{p} = 0$, orbital angular momentum has no component along its direction of motion, so helicity is equal to the component of the *total* angular momentum along that direction.

Chapter 12

[1] The correct formula for the total (rest plus kinetic) energy of a particle of mass m and velocity v, according to Einstein's

special relativity, is $mc^2/\sqrt{1-(v/c)^2}$. For small v/c, this expression is approximately equal to $mc^2 + mv^2/2$, which is the rest energy plus the non-relativistic kinetic energy. For v approaching c, we can see from this formula that the energy goes to infinity.

Chapter 13

[1] This is just an order of magnitude estimate often used in this book, not an exact calculation. For a more accurate estimate, we should have used $3kT/2$ rather than kT as the kinetic energy of a particle, but even so it is still an approximation because that is just the average of a complicated distribution of energies.

[2] This is known as the *Stefan–Boltzmann law.* See note 16[1] for a derivation using a dimensional argument. Such a derivation makes it clear that these proportionalities are also true for any relativistic gas where the rest mass is zero or negligible.

[3] We start from the *First Law of Thermodynamics,* which is simply the statement that in the absence of any heat exchange, the energy lost in a volume must be equal to the work done to expand this volume. Thus, if E is the energy contained in a volume V, then an infinitesimal change of E and V are related by $dE = -p\, dV$, where p is the pressure exerted by the surroundings on the volume to prevent its expansion.

The energy E of relativistic particles contained in a volume V is ρV, where ρ is the energy density, and the pressure is $p = \rho/3$. Substituting this into the First Law of Thermodynamics, and using the Stefan–Boltzmann law, which says that ρ is proportional to T^4, we obtain $4V\, dT + T\, dV = -T\, dV/3$ — hence, $3dT/T = -dV/V$. This demands T^3 to be inversely proportional to V, or equivalently, T to be inversely proportional to the linear size of the volume.

[4] The universe is everything; there is nothing for it to exchange heat with, so the result obtained in the last note is applicable to the universe if it contained only relativistic particles. A more sophisticated way of saying that for people who know the Second Law of Thermodynamics is that the universe is uniform and homogeneous, so its entropy cannot change because it is already at its maximum value.

However, the universe contains non-relativistic as well as relativistic particles. Thus, in applying the First Law of Thermodynamics, we should take the energy to be the sum of the relativistic energy E and the non-relativistic energy E', and the pressure to be the sum of the relativistic pressure p and the non-relativistic pressure p'. Since $p' = 0$ and E' is constant (matter conservation law tells us that the number of non-relativistic particles is constant, and the energy per particle is dominated by the rest energy which is also constant), we have $dE' = -p'\,dV = 0$, hence the First Law of Thermodynamics can indeed be written $dE = -p\,dV$, as in note 13[3]. As shown in that note, the temperature of the relativistic particles is inversely proportional to a. Since in thermal equilibrium all particles in the volume have the same temperature, this relation is true for non-relativistic particles as well — hence, the whole universe.

Chapter 14

[1] The curve is taken from J. C. Mather *et al.*, *Astrophysical Journal* **420** (1994) 439.

[2] The 3-year WMAP data shown are taken from G. Hinshaw *et al.*, http://arxiv.org/pdf/astro-ph/0603451.

Chapter 15

[1] From the First Law of Thermodynamics (note 13[3]), $d(\rho V) = -p\, dV$, we see that if ρ is a constant then $p = -\rho$, giving rise to a negative pressure which sucks the universe outward to cause an acceleration. More generally, if ρ is proportional to the $(-w - 1)$th power of V, then $p = w\rho$.

It can be shown that the universe accelerates or decelerates depending on whether the quantity $\rho + 3p$ is negative or positive. With dark energy alone, whose pressure is given by $p = w\rho$, this sum is $(3w + 1)\rho$. With 73% dark energy and 27% non-relativistic matter which provides no pressure, this sum is $(3w + 1)\rho_{de} + \rho_m = (3w + 1.37)\rho_{de}$, where ρ_{de} and ρ_m are the energy densities of dark energy and matter, respectively. We know this number to be negative because the present universe is accelerating. This requires $w < -0.36$. A careful analysis of WMAP and other data puts w much closer to, and in fact to be completely consistent with, –1, the value it should have if dark energy is the vacuum energy.

[2] The energy density of non-relativistic matter is proportional to $1/a^3$, hence $(z + 1)^3$, and the energy density of dark matter is constant. The present ratio of dark energy and non-relativistic matter energy densities is 73% to 27%, so at any redshift z their ratio would be 73 to $27(z + 1)^3$. Thus, these two kinds of energy densities become equal at $z + 1 = (73/27)^{1/3} = 1.39$, giving rise to $z = 0.39$.

[3] Consider a test object with mass m placed at a distance r not far from us, as shown in the following figure. If ρ is the *energy* density of the universe, then the *mass* inside a sphere of radius r (colored yellow) is $M = 4\pi\rho r^3/3c^2$. This sphere expands with the universe, causing the object with mass m to move outward

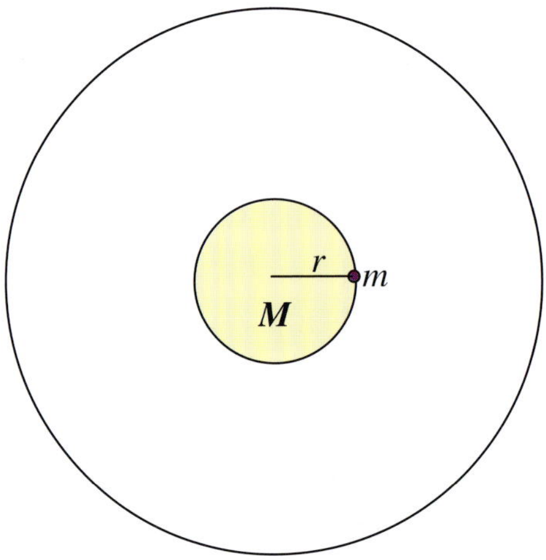

Figure 53: This diagram is used to derive the Friedmann equation.

with it at the same rate. As it is situated not far from us, its velocity dr/dt is small and non-relativistic.

The total energy of the test object is the sum of its rest energy mc^2, its potential energy due to the gravitational attraction of all the galaxies in the universe, which is (Chap. 12) $PE = -GMm/r$, and its kinetic energy $KE = m(dr/dt)^2/2$ (Chap. 12). In reaching this conclusion, we have made use of a classical result that the gravitational force and potential on the object in a uniform universe come only from those galaxies inside the yellow sphere. The attraction from the rest cancels one another out.

Since the rest energy is independent of time, so must be the sum $KE + PE$. If x is the comoving (i.e. present) distance of the object, and $a(t)$ is the scale factor at time t, then $r = a(t)x$ and x is independent of time. If we define a constant k so that the sum $KE + PE$ is equal to $-kmx^2/2$, and substitute in the explicit expressions for KE, PE, and M, we obtain the Friedmann equation $(da/dt)^2 - 8\pi\rho Ga^2/3c^2 = -k$. This equation does not involve either m or x, so it is an equation describing the evolution of the universe, independent of the test object we used to derive it. The input of this equation is the density ρ of the universe, expressed as a function of the scale factor a. What that is depends on the nature of the constituent in the universe at that time.

If it is non-relativistic matter, the total energy is conserved so ρ is proportional to a^{-3}. If it is relativistic particles, then ρ is proportional to T^4 and hence a^{-4}. If it is dark energy, or during inflation, ρ is a constant independent of a.

Since the total energy of the universe is proportional to ρa^3, other than the first case, the total energy varies with a. In these cases, a pressure p is present, and the energy gain or loss is attributed to the work done to overcome this pressure p. The equation expressing this, $(4\pi/3)d(\rho a^3) = -p(4\pi a^2\, da)$, is known as the *continuity equation*. For relativistic particles, this implies $p = \rho/3$, as we know. For dark energy or inflation energy, it is $p = -\rho$.

[4] Since $a(da/dt)^2 = (4/9)(da^{3/2}/dt)^2$ is a constant, $a^{3/2}$ must be proportional to t, and hence a must be proportional to $t^{2/3}$. The Hubble constant H_0 at the present time t_0 is therefore $H_0 = (da/dt)/a = (2/3)t_0$, and hence $t_0 = (2/3)(1/H_0)$.

The radiation era is very short compared to the matter era, so in computing the age of the universe we may ignore the presence of the radiation era. For that reason, t_0 is just the age of the universe.

[5] The present fraction of critical density is usually denoted by Ω. Thus, $\Omega_{\rm m} = 0.27$ is the matter fraction, and $\Omega_{\rm de} = 0.73$ is the dark energy fraction. The photon fraction can be computed from the Stefan–Boltzmann law in note 18[3] to be $\Omega_\gamma = 4.8 \times 10^{-5}$ at the present temperature of 2.725 K. If we assume the three species of neutrino to be massless, and

take into account their present temperature to be 1.9 K (see Chap. 19), then the fraction of total radiation density including photons and neutrinos is $\Omega_r = 8.2 \times 10^{-5}$.

Let us assume the dark energy density to be constant in time. Radiation energy density is proportional to $1/a^4$, and matter energy density is proportional to $1/a^3$. Hence, the energy density for a flat universe at any time is given in terms of the critical density ρ_{crit} to be $\rho = \rho_{crit}(\Omega_{de} + \Omega_m/a^3 + \Omega_r/a^4)$. With that, and the relation $H_0 = (8\pi\rho_{crit}G/3c^2)^{1/2}$, the Friedmann equation becomes $da/dt = H_0\, a(\Omega_{de} + \Omega_m/a^3 + \Omega_r/a^4)^{1/2}$. With the initial condition $a(0) = 0$, this equation can be solved numerically at any moment to obtain a relation between the physical time t and the scale factor a.

[6] The solution $a = a_0 \exp(\alpha t)$ is the solution of Friedmann equation for $k = 0$, where $\alpha = (8\pi G\rho/3c^2)^{1/2}$ is also its Hubble parameter H, and a_0 is the initial size of the universe at $t = 0$. If the universe is not critical, then this exponential growth is true at late time when k becomes insignificant compared to either term on the left-hand side of the Friedmann equation.

Chapter 16

[1] The values for these three fundamental constants are: $c = 3 \times 10^8$ m/s, $\hbar = 1.05 \times 10^{-34}$ kg m^2/s, and $G = 6.67 \times 10^{-11}$ m^3/kg/s^2. After converting to electronvolts, the product $\hbar c$ becomes 1.97×10^{-7} eV m. The combination $M_P = (\hbar c/G)^{1/2} = 2.18 \times 10^{-8}$ kg is the fundamental unit of mass, called the Planck mass. In terms of its rest energy, it is $M_P c^2 = 1.22 \times 10^{19}$ GeV. The combination $\ell_P = (\hbar G/c^3)^{1/2} = 1.6 \times 10^{-35}$ m is the fundamental unit of length, called the Planck length, and the combination $t_P = (\hbar G/c^5)^{1/2} = 5.4 \times 10^{-44}$ seconds is the fundamental unit of time, called the Planck time.

Incidentally, the Stefan–Boltzmann law discussed in Chap. 13 says that the energy density ρ of photon is proportional to T^4, the fourth power of the temperature, and its number density is proportional to T^3. These relations can be obtained as follows by a 'dimensional analysis,' similar to that used in the last paragraph.

To obtain an expression for ρ, we must first determine what it can depend on.

Since we are dealing with photons, we expect $\hbar$ and c to be involved but not G, because gravity is irrelevant in this discussion. The Boltzmann constant k may enter as well because we are dealing with a thermal problem. The only other quantity that ρ can depend on is the temperature T.

Now the only combination of $\hbar$, c, k, and T that has the unit ('dimension') of energy density is $(kT)^4/(\hbar c)^3$; hence, the energy density ρ must be proportional to this quantity. The proportionality constant for a photon gas can be obtained only by detailed calculations and it turns out to be $\pi^2/15$. In the same vein, the only combination that can yield a unit of number density is $(kT)^3/(\hbar c)^3$, so the photon number density must be proportional to that. The proportionality constant for photons turns out to be $2.404/\pi^2$. In particular, at the present temperature of 2.725 K, the density of photons is $4.12 \times 10^8/\text{m}^3$. In contrast, since the energy density of nucleons is 4.5% of the critical density of 0.97×10^{-26} kilograms per cubic meter, and the mass of each nucleon is 1.7×10^{-27} kg, the nucleon number density is about 0.25 per m^3. The ratio of the nucleon to photon density is usually called η. It is a very small number, less than 10^{-9}.

[2] See note 15[6].

[3] As shown in note 9[2], the comoving distance travelled by light after a time t is given by the integral $\int c\, dt'/a(t')$,

integrated from 0 to t. The physical distance traveled is then equal to $a(t)$ times this integral.

In the radiation era, $a(t)$ is proportional to $t^{1/2}$; hence, this physical distance is $2ct$. It is larger than ct because the expansion of the universe carries light further along than a static universe, much like a boat going downstream travels faster than it does in still water. In the matter era, $a(t)$ is proportional to $t^{2/3}$; hence, this physical distance is $3ct$. It travels farther than in the radiation era because with the absence of pressure the universe expands even faster.

In the inflationary era, $a(t) = a_0 \exp(\alpha t)$, the physical distance becomes $(c/\alpha)\exp(\alpha t)$ when the exponential factor is much larger than 1. Light can reach the whole universe to make it homogeneous as long as this distance is larger than $a(t)$, namely, when $c/\alpha > a_0$. In words, this happens when light can travel across the initial region within the characteristic time $1/\alpha$. This is the time during which the universe has inflated $e = 2.71828$ times its original size.

[4] Let S be the seed energy before inflation. Then the total energy at the beginning of the classical Big Bang is SF^3. At that time the temperature is T_{init}. Since the energy of relativistic particles in the universe is proportional to T, by the time the universe enters from the radiation era into the matter era at temperature T_{eq}, the total relativistic energy has reduced to $SF^3(T_{\text{eq}}/T_{\text{init}})$. But this is the time when non-relativistic matter energy is equal to the relativistic photon and neutrino energy, so the amount of non-relativistic energy at that time is also given by this amount. Since the total amount of relativistic energy does not change with time, this is also the present amount of non-relativistic energy. How much is that?

At the end of Chap. 10, it was mentioned that the amount of ordinary matter in the observable universe today is estimated to be 7.5×10^{53} kilograms. Moreover, there is about five times as much dark matter, making a total mass of about 4.5×10^{54} kilograms. This then is the total amount of non-relativistic energy in the observable universe. Equating the two, we get $SF^3(T_{\text{eq}}/T_{\text{init}}) = 4.5 \times 10^{54}$ kilograms; hence, $S = 4.5 \times 10^{54} F^{-3} (T_{\text{init}}/T_{\text{eq}})$ kilograms $= 4.5 \times 10^{54} F^{-3} (T_{\text{init}}/0.74 \text{ eV})$ kg.

[5] I am grateful to James Bjorken for emphasizing this point to me.

Chapter 17

[1] We encountered this kind of stretching of light waves although we might not have realized it. The CMB left its source at a decoupling temperature T_* of about ¼ eV, reaching us today when the temperature of the universe is $T_0 = 2.725$ K. Since the wavelength is inversely proportional to its energy and hence the temperature, between then and now it must have increased a factor $T_*/T_0 = a_0/a_*$, which is precisely how much the size of the universe has increased. Thus, the photon wavelength just stretches with the universe, as our intuition tells us to be the case. Sound waves do the same thing.

[2] The local relations between the physical distance r and the comoving distance x is $dr = a(t)\ dx$. The local relation between the physical time t and the conformal time η is $dt = a(t)\ d\eta$. Hence, speed in both coordinate systems is the same because $dr/dt = dx/d\eta$.

[3] H rather than da/dt is the appropriate parameter to measure the expansion rate with because the latter depends on how a is normalized, and physical effects should not depend on normalization conventions. Quantitative calculation supports this assertion. Throughout this book we normalize a to be 1 at the present time but that is just a convenient choice, not inherent in the physics.

The distance and time in the comoving frame are obtained, respectively, from the physical distance and time by a division by *a,* so quantities with a dimension (unit) of an inverse time, which is what *H* is, should be multiplied by a factor *a* to obtain their equivalent in the comoving frame.

[4] This graph is taken from M. Tegmark *et al.*, *Astrophysical Journal* **606** (2004) 702.

[5] From the note 16[1], $\hbar c = 1.97 \times 10^{-7}$ eV m. Thus, $V = \rho(\hbar c)^3$ has a dimension of energy to the fourth power.

[6] For example, if $\varepsilon = 5.3 \times 10^{-4}$, then $V^{1/4} = 10^{16}$ GeV, and the exponent for inflation α is 3.6×10^{37} s^{-1}. Thus, the time *t* it takes for $\exp(\alpha t)$ to grow $10^{25} = \exp(57.6)$ times is 1.6×10^{-36} seconds, a very short time indeed by ordinary standards. However, this time is very long in some sense because it has allowed the universe to undergo 57.6 'e-foldings' before the false vacuum decays away.

One might ask whether the slow-roll parameter ε is slow enough to allow so many e-foldings. The answer depends on the details of the false vacuum potential, as illustrated by the following example.

Suppose the false vacuum is given by the potential $V(\phi) = m^2\phi^2/2$, where ϕ tells us where the *inflaton* is in energy space, and *m* is a parameter with dimension of energy. If the false vacuum is located at $\phi = \phi_0$ and the true vacuum at $\phi = 0$, then the parameter *V* in the text is simply $V = V(\phi_0)$. For any false vacuum potential $V(\phi)$, the slow-roll parameter ε is defined to be $\varepsilon = (M_P^2/16\pi)(dV/d\phi)^2/V(\phi)^2$, evaluated at $\phi = \phi_0$. For the potential in hand, this becomes $\varepsilon = (M_P^2/4\pi\phi_0^2) = (M_P^2 m^2/8\pi V)$. Putting in the values of ε and *V* in the text, we get $m^2 = 8\pi V\varepsilon/M_P^2 = (9.5\times10^{11}\ \text{GeV})^2$. Hence, $\phi_0^2 = 2V/m^2 = (1.5\times10^{20}\ \text{GeV})^2 > M_P^2$. Some find that objectionable because without quantum gravity, we should not trust anything whose energy is greater than the Planck mass.

[7] See note 14[2] for the reference from which this graph is taken.

[8] Consider Fig. 54, marking out the various conformal distances of the frozen universe, with us at the center, the decoupling surface and the Big Bang surface represented by the solid and the dashed circles, respectively. These are the surfaces where light emanating at the times of decoupling and Big Bang, respectively, have just reached us. Thus, if η_0 is the conformal age of the universe, η_* the conformal time at decoupling, and c the speed of light, then the comoving distances of these two surfaces to us are, respectively, $c(\eta_0 - \eta_*)$ and $c\eta_0\theta$. However, since η_0/η_* is about 50,[9] we can approximate $c(\eta_0 - \eta_*)$ by $c\eta_0\theta$. From the blue curve in Fig. 45 we know that is $\lambda/2$. Equating these two, we get $\lambda = 2c\eta_0\theta$, so the relation between k, l, and θ is $k = 2\pi/\lambda = \pi/c\eta_0\theta = l/c\eta_0$.

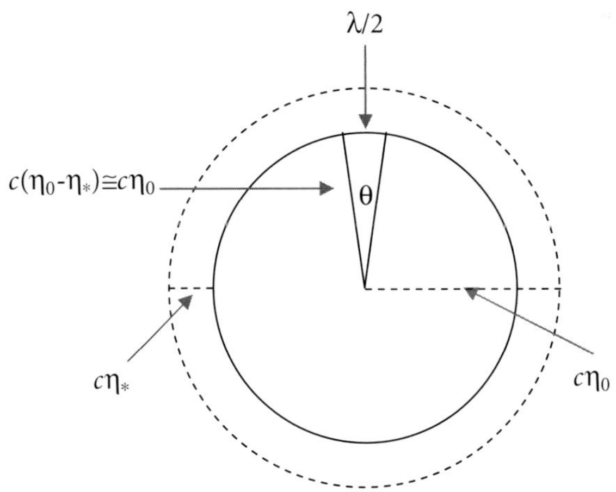

Figure 54: In this picture of the universe, the earth is at the center, the dashed circle is the surface of the Big Bang, and the solid angle is the surface of decoupling. The geometry of the universe is assumed to be flat.

[9] The conformal time at various epochs can be obtained by solving the Friedmann equation $da/d\eta = H_0 a^2(\Omega_{\rm de} + \Omega_{\rm m}/a^3 + \Omega_{\rm r}/a^4)^{1/2}$ obtained from note 15[5] by substituting $dt = a\ d\eta$. To compute η_0, the condition $a(\eta_0) = 1$ has to be used. To compute η_*, the condition $a(\eta_*) = 1/(z_* + 1) = 1090$ is used, and to compute $\eta_{\rm eq}$ when matter and radiation energies are equal, the condition $a(\eta_{\rm eq}) = \Omega_{\rm m}/\Omega_{\rm de}$ is used. In this way, we obtain $\eta_0 = 4.7 \times 10^{10}$ years, $\eta_* = 9.2 \times 10^8$ years, and $\eta_{\rm eq} = 3.9 \times 10^8$ years. These conformal times are bigger than the corresponding physical times because the scale factor is less than 1. In units of Mpc/h, the conformal distance to the Big Bang surface is $c\eta_0 = 1.04 \times 10^4$ Mpc/h, and the sound horizon is $c_{\rm s}\eta_* = c\eta_*/\sqrt{3} = 117$ Mpc/h. The first acoustic peak in Fig. 44 therefore occurs at a wave number $k = l/c\eta_0 = 2.1 \times 10^{-2}$ h/Mpc.

[10] The red curve in Fig. 45 is the wave displacement $A\cos(2\pi\eta/T)$, plotted against the wave number k, with the amplitude A set equal to 1. The conformal period of oscillation is (see the discussion around Fig. 16) $T = \lambda/c_{\rm s} = 2\pi\sqrt{3}/kc$. At the time of decoupling, $\eta = \eta_*$, the peaks of the cosine function are located at integral multiples of η_*/T. In the variable k, the peaks of the red curve are separated by a wave number $\Delta k = 2\pi\sqrt{3}/c\eta_*$, and the peaks in the blue curve of displacement square are separated by a wave number $\Delta k = \pi\sqrt{3}/c\eta_*$.

[11] Suppose the universe has a positive curvature like the sphere shown in Fig. 21. Take the curved triangle made by the three red apices and map it onto the triangle in Fig. 54 so that the north pole is mapped onto the center of the circle, and the other two red apices are mapped onto the other two vertices of the flat triangle. Then the curved triangle bulges outside the flat triangle of Fig. 54, making the angle subtended by

the curved line at the north pole larger than the angle θ of the flat triangle. Hence, the multiple $l = \pi/\theta$ of the first peak is smaller in the positive-curvature space than the flat space. For a space with negative curvature, the two geodesic lines curve the other way, consequently it leads to a smaller θ and a large l compared to the flat space.

[12] According to note 15[5], the energy density of matter is proportional to Ω_m/a^3, and the energy of radiation is proportional to Ω_r/a^4. These two become equal when $a = \Omega_r/\Omega_m$.

By definition, aH is equal to da/dt. Using the Friedmann equation at the end of note 15[5], we can calculate $aH/c = da/dt/c$ at $a = \Omega_r/\Omega_m$.The result is 0.014 h/Mpc.

[13] D. J. Eisenstein *et al.*, *Astrophysical Journal* **633** (2005) 560.

Chapter 18

[1] According to the special theory of relativity, the total energy of a particle with mass m and velocity v is $mc^2/\sqrt{1-v^2/c^2}$. When $v = c$, this remains a finite number only when $m = 0$. If $v < c$, then we must have $m > 0$ in order for this total energy to remain positive.

[2] See note 11[2].

[3] According note 16[1], the number density of photons is given by the formula $1.202\pi^2 g(kT/\hbar c)^3$, and the energy density of photons is given by the formula $(\pi^2 g/30)(kT)^4/(\hbar c)^3$, where $g = 2$ is the number of allowed helicities of the photon. These formulas are correct for every relativistic boson, and with an additional factor of 3/4 for the first formula and a factor 7/8 for the second, they are also correct for every relativistic fermion and very relativistic anti-fermion.

The total energy density in the radiation era is therefore $\rho = (\pi^2 g_*/30)(kT)^4/(\hbar c)^3$, where g_* is the effective number of degrees of freedom, counting 1 for each helicity of each boson, and 7/8 for each helicity of each fermion.

Similarly, the total number density in the radiation era is $n = 1.202\pi^2 g'(kT/\hbar c)^3$, where g' is obtained from the sum of bosons and fermions, counting 1 for each helicity of each boson, and ¾ for each helicity of each fermion.

The Hubble parameter H in the radiation era can be calculated using this relation of ρ. In particular, early in the radiation era, all SM particles are relativistic. Since each quark comes in three *colors,* the effective number of degrees of freedom for all the SM particles can be shown to be $g_* = 106.75$. The Hubble expansion rate early in the radiation era is therefore

$$H = \sqrt{8\pi G\rho/3} = \sqrt{8\pi^3/90 g_*}\, T^2/M_p = 1.66\sqrt{g_*}\, T^2/M_p,$$

which works out to be about 17 T^2/M_P.

[4] Assuming the masses of the leptons ℓ and the Higgs ϕ to be much smaller than the mass M_1 of the heavy Majorana fermion N_1, the only parameters the decay rate of the process $N_1 \to \ell + \phi$ can depend on are the mass M_1 and the dimensionless coupling s_1. Of course the decay rate may depend on fundamental physical constants like c and $\hbar$ as well. The only way to use these parameters to construct a quantity with a dimension of an inverse time that decays is to have the combination $M_1 c^2/\hbar$. Hence, the decay rate must be equal to known constant times $s_1 M_1 c^2/\hbar$, or equivalently, some other known constant times $s_1 M_1$.

Chapter 19

[1] There is a quantity called entropy given by $S = (\rho + p)V/T = 4\rho V/3T$ for relativistic particles. This quantity is supposed to

be conserved for our universe. Using the Stefan–Boltzmann law (see note 18[3]), the energy density ρ in the radiation era is equal to $(\pi^2/30)g_*(kT)^4/(\hbar c)^3$, where g_* is the effective degree of freedom of relativistic particles, counting 1 for every helicity state of every boson, and 7/8 for every helicity state of every fermion. After 100 MeV or so, the only relativistic particles left are the photons, the electrons, the positrons, the neutrinos and the anti-neutrinos of all three generations. This yields a $g_* = 2 + (7/8)(2 + 2 + 3 + 3) = 43/4$.

After neutrino decoupling, the effective number of degrees of freedom is reduced to $g_* = 2 + (7/8)(2 + 2) = 11/2$. After electron–positron annihilation, this number is further reduced to 2. Since energy density cannot suddenly change, this sudden reduction of g_* by a factor 11/4 must be accompanied by an increase in temperature by a factor $(11/4)^{1/3} = 1.40$.

[2] In the radiation era, the scale factor a is proportional to the square root of time t, so the Hubble parameter H is equal to $1/2t$. According to the Friedmann equation, H^2 is equal to $8\pi G\rho/3c^2$, which is proportional to T^4. Using the value of $g_* = 43/4$, this gives a relation between t and T to be $tT^2 = 0.78$ $(\mathrm{MeV})^2$ seconds. Using this formula, the time it takes to reach a temperature of 0.1 MeV is 78 seconds. The actual time is a bit longer because after electron–positron annihilation, $g_* = 2$, so $tT^2 = 1.8$ $(\mathrm{MeV})^2$ seconds.

BBN is over when the temperature gets down to about 30 keV. Using the second formula, this corresponds to a time of 33 minutes.

[3] If M is the mass of the dead star, and m the mass of an object placed a distance r from the center of the star, then the gravitational potential energy between the two is $-GMm/r$ (Chap. 12). If this is larger than the rest energy mc^2 in magnitude, then the total energy is negative and the

gravitational hold on the mass m is so strong that it can never escape to infinity, where the potential energy is zero and the total energy positive. This would happen when the distance r is less than GM/c^2, independent of m. However, this Newtonian estimate is not completely reliable because we need Einstein's general relativity to deal with the strong gravitational force encountered in this problem. Using general relativity, the result becomes $r = 2GM/c^2$. Nothing within this distance can escape its gravitational hold on the dead star, not even light. Consequently, the dead star appears to be black, and is thus known as a black hole. The distance r given above is called the black hole horizon.

[4] We encountered the Pauli exclusion principle in Chap. 11, used there to reduce the effective attraction of the nucleons added in later. A more precise description of the principle will be given below, together with its relevance to the Chandrasekhar limit. For simplicity of description, I shall pretend we live in a one-dimensional space. The generalization to three dimensions is straightforward mathematically, but it is a bit awkward to describe in words, which is why I am confining myself to one dimension.

Consider a number of fermions of a particular kind, say electrons, put into a box of length L. Like everything else, they want to settle down to a configuration of minimal energy. We want to know what that configuration is.

Assuming they do not interact with one another, then the total energy is the sum of their individual kinetic energies $p^2/2m$, where m is the mass of the fermion and p is its momentum. Ignoring quantum mechanics, we know what to do. To minimize the energy we simply let each of them to have $p = 0$. With quantum mechanics, we are not allowed to do that because of the uncertainty principle.

Moreover, there is also the Pauli exclusion principle, which demands that no two fermions of the same helicity be allowed to be at the same place with the same momentum.

By 'the same place' and 'the same momentum,' I mean the following. Draw a rectangle of width L and an unspecified height (Fig. 55). Label the base by the position x and the height from the bottom by the momentum p of a fermion. Next, divide the rectangle into many smaller but identical boxes, each having an area $h = 2\pi\hbar$. It does not matter what is the width or the height of these smaller boxes, as long as they have the same shape and have this same area h. The position and momentum of a fermion (black dots) can now be determined by which box it is in. Note that there is an uncertainty as to the precise position and the precise momentum in the box, agreeing with the quantum mechanical uncertainty principle.

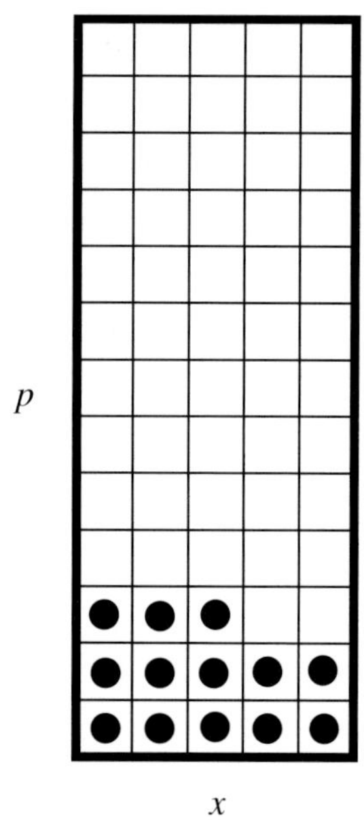

Figure 55: A diagram illustrating the Pauli exclusion principle in one dimension.

The Pauli Exclusion Principle specifies that no two fermions of the same helicity may occupy the same box. Assuming the black dots in Fig. 55 all have the same helicity, then with 13 fermions what is shown in Fig. 55 is one way to arrange them to minimize the total fermion energy.

If we add more fermions, then the late-comers have to occupy higher layers of the rectangle; hence, they have larger momentum and larger kinetic energy. If these are nucleons attracted to other nucleons with a negative potential energy, this inevitable kinetic energy is going to dilute the negative potential energy and reduce the effective interaction. This is how the Pauli exclusion principle was expressed in Chap. 11.

When the exclusion principle is applied to a dead star, we can imagine ℓ to be the diameter of the star. When gravity acts to compress the star, changing its size from ℓ to $4\ell/5$, for example, then in Fig. 55, each layer can only accommodate four boxes instead of five. The displaced boxes must now move to the top, and in so doing energy must be supplied. If the released gravitational energy due to contraction is insufficient to supply the energy needed for this move, the contraction will stop. This is essentially now the electrons hold up a dwarf and neutrons hold up a neutron star to prevent them from further gravitational contraction.

Appendix B: Abbreviations and Mathematical Symbols

Symbols	Meaning	Page
2dFGRS	2-degree Field Galaxy Redshift Survey	114
α	inflation exponent	210, 212, 214
a	scale factor	58, 110
a_*	scale factor at decoupling	112
a_{eq}	scale factor at radiation-matter equality	111
b	bottom quark	168
B	baryonic number	159
BBN	Big Bang Nucleosynthesis	188
γ	photon	164
c	speed of light	37, 86, 124, 210
c	charm quark	168
c_s	speed of sound	41, 151
C	Celsius scale of temperature	97
C	charge conjugation operation	165
CMB	cosmic microwave background radiation	103
COBE	Cosmic Background Explorer	104
d	down quark	162

Symbols	Meaning	Page
d	distance	46
ε	slow-roll parameter	147
e	electron	68, 163
eV	electronvolt	89
E	energy	86
ϕ	Higgs Boson	166, 169, 181
g_*	effective number of degrees of freedom	217
G	Newtonian gravitational constant	85, 124, 210
GeV	Giga-electronvolt = 10^9 eV	89
g	gluon	164
η	slow-roll parameter	147
η	ratio of number of nucleons to number of photons	159, 180, 189, 191
η	conformal time	142
η_*	conformal time at decoupling	151
η_0	conformal age of the universe	149
h	$2\pi\hbar$	221
h	H_0 in units of 100 km/s/Mpc	137, 150
$\hbar$	Planck's constant	124, 140, 210
H	Hubble parameter	117
H_0	Hubble constant at the present time	46
k	Boltzmann constant	98
k	the constant on the right of the Friedmann equation	116
k	wave number	138, 149
keV	kilo-electronvolt = 10^3 eV	89
km	kilometer = 10^3 m	47
K	Kelvin scale of temperature	97
KE	kinetic energy	83
λ	wavelength	138, 203, 215

Symbols	Meaning	Page
ℓ_p	Planck length	210
l	CMB multiple moment	148
L	leptonic number	159
μ	muon	163
m	mass	83, 86
meV	milli-electronvolt = 10^{-3} eV	89
MeV	mega-electronvolt = 10^{6} eV	89
M_P	Planck mass	210
Mpc	megaparsec = 10^{6} pc	39
ν	(light) neutrino (left-handed)	75, 163
n	tilt of CMB power spectrum	147
n	neutron	71, 192
nm	nanometer = 10^{-9} m	100
N	(heavy) neutrino (right-handed)	172, 180
p	pressure	205, 207, 209
p	proton	69, 192
pc	parsec = 3.26 light years	39
P	parity operation	165
PE	Potential energy	83
θ	CMB correlation angle	148
ρ_{crit}	critical energy density	64, 117
ρ	energy density	98, 116, 217
s	strange quark	168
s	second	47
SDSS	Sloane Digital Sky Survey	113
SM	Standard Model of Particle Physics	161
τ	tau particle	168
t	time	58
t	top quark	168
t_0	present age of the universe	110
t_P	Planck time	210

Symbols	Meaning	Page
T	time-reversal operation	165
T	temperature	98, 110
T_*	temperature at decoupling	111
T_0	temperature of the universe at the present time	110
T_{eq}	temperature at radiation-matter equality	111
T_{init}	temperature at reheating	129
TeV	tera-electronvolt = 10^{12} eV	89
u	up quark	162
v	velocity	46, 83
V	volume	203
$W^{\pm}$	W boson	164
WMAP	Wilkinson Microwave Anisotropy Probe	104
z	redshift	46, 60, 110
z_*	redshift at decoupling	111
z_{de}	redshift at matter-dark energy equality	114
z_{eq}	redshift at radiation-matter equality	111
z_{r}	redshift at re-ionization	113
Z	Z boson	164

Appendix C: Important Events after Reheating

T(eV)	T(K)	z	a	t (years)	η (years)	Event	Page
$> 10^9$	$>10^{13}$	$>10^{13}$	$<10^{-13}$	0	0	leptogenesis	180
$\sim 10^6$ -3×10^4	$\sim 10^{10}$ -10^8	$\sim 10^{10}$ -10^8	$\sim 10^{-10}$ -10^{-8}	~1–2000 seconds	1.5×10^3	ν decoupling, e^+e^- annihilation, BBN	189
0.74	8600	3233	3.1×10^{-4}	5.7×10^4	3.9×10^8	radiation-matter equality	111
0.26	2970	1089	9.2×10^{-4}	3.8×10^5	9.2×10^8	decoupling	111
~0.005	~57	~20	~0.048	$\sim 1.8 \times 10^8$	$\sim 1.1 \times 10^{10}$	star formation	113
0.00033	3.8	0.39	0.72	9.5×10^9	4.2×10^{10}	matter-dark energy equality	114
0.00024	2.725	0	1	1.37×10^{10}	4.7×10^{10}	present time	48, 103, 110, 216

Index